Battle
& Te

£6.50

SMALL
ARMS
&
CANNONS

Other Titles in the
Battlefield Weapons Systems and Technology Series

General Editor: Colonel R G Lee OBE, Military Director of Studies at the Royal
Military College of Science, Shrivenham, UK

This new series of course manuals is written by senior lecturing staff at RMCS,
Shrivenham, one of the world's foremost institutions for military science and its
application. It provides a clear and concise survey of the complex systems spectrum of
modern ground warfare for officers-in-training and volunteer reserves throughout the
English-speaking world.

For full details of these and future titles in the series, please contact your local
Brassey's/Pergamon office

Related Titles of Interest

SMALL ARMS & CANNONS

C. J. Marchant Smith

and

P. R. Haslam

Royal Military College of Science, Shrivenham, U.K.

BRASSEY'S PUBLISHERS LIMITED
a member of the Pergamon Group

OXFORD · NEW YORK · TORONTO
SYDNEY · PARIS · FRANKFURT

U.K.	BRASSEY'S PUBLISHERS LIMITED, a member of the Pergamon Group, Headington Hill Hall, Oxford OX3 0BW, England
U.S.A.	Pergamon Press Inc., Maxwell House, Fairview Park, Elmsford, New York 10523, U.S.A.
CANADA	Pergamon Press Canada Ltd., Suite 104, 150 Consumers Rd., Willowdale, Ontario M2J 1P9, Canada
AUSTRALIA	Pergamon Press (Aust.) Pty. Ltd., P.O. Box 544, Potts Point, N.S.W. 2011, Australia
FRANCE	Pergamon Press SARL, 24 rue des Ecoles, 75240 Paris, Cedex 05, France
FEDERAL REPUBLIC OF GERMANY	Pergamon Press GmbH, 6242 Kronberg-Taunus, Hammerweg 6, Federal Republic of Germany

First edition 1982

Library of Congress Cataloging in Publication Data

Marchant Smith, C. J.
Small arms & cannons.
(Battlefield weapons systems & technology ; v. 5)
Includes index.
1. Firearms. 2. Ordnance. I. Haslam, P. R.
II. Title. III. Title: Small arms and cannons.
V. Series.
UD380.M37   1982   623.4'4   82-3801

British Library Cataloguing in Publication Data

Marchant Smith, C. J.
Small arms & cannons.—(Battlefield weapons
systems and technology; v. 5)
1. Firearms
I. Title II. Haslam, P. R. III. Series
623.4'4   UF520

ISBN 0-08-028330-6 (Hardcover)
ISBN 0-08-028331-4 (Flexicover)

In order to make this volume available as economically and as rapidly as possible the authors' typescripts have been reproduced in their original forms. This method unfortunately has its typographical limitations but it is hoped that they in no way distract the reader.

The views expressed in the book are those of the authors and not necessarily those of the Ministry of Defence of the United Kingdom.

Printed in Great Britain by A. Wheaton & Co. Ltd., Exeter

Preface

The Series

This series of books is written for those who wish to improve their knowledge of military weapons and equipment. It is equally relevant to professional soldiers, those involved in developing or producing military weapons or indeed anyone interested in the art of modern warfare.

All the texts are written in a way which assumes no mathematical knowledge and no more technical depth than would be gleaned from school days. It is intended that the books should be of particular interest to army officers who are studying for promotion examinations, furthering their knowledge at specialist arms schools or attending command and staff schools.

The authors of the books are all members of the staff of the Royal Military College of Science, Shrivenham, which is comprised of a unique blend of academic and military experts. They are not only leaders in the technology of their subjects, but are aware of what the military practitioner needs to know. It is difficult to imagine any group of persons more fitted to write about the application of technology to the battlefield.

Volume V

This book aims to explain the requirements for small arms and cannon, and in outline how the designer meets those requirements. Whilst most books on small arms are a descriptive catalogue of types, this approach is not followed here, weapons are only named or pictured if they illustrate a point. Apart from giving the professional soldier the sort of understanding of his personal weapon which the authors believe he should have, it may also be useful to others who wish to enquire rather more into the design philosophy of small arms and cannon.

Shrivenham, November 1981   Geoffrey Lee

Acknowledgements

The authors greatly appreciate the help they have received
from members of many Service Establishments.   In
particular from the staffs of the Royal Small Arms Factory
Enfield, the Mechanical Design and the Applied
Thermodynamics Branches of the Royal Military College
of Science.

The authors are also most grateful to the Photographic
and Ballistics Departments at RMCS for assistance with
the photographs and to Mr. J. F. Parker for all his work
on the diagrams.

Bill Marchant Smith
Paul Haslam

Contents

List of Tables

List of Illustrations

Chapter 8

1.

Outline Military Requirements

Fundamentals

The layman's view of small arms and ammunition is often a romantic notion of
the capabilities of fabled guns in the hands of legendary characters: sometimes it
is an awareness of those weapons used by guerrillas or it may be something
vaguely called a 'rifle' which is carried by a soldier. Apart from may be hearing
the calibre as part of the name of some weapons, he is unlikely to take any particular note of the ammunition. It does little to catch the imagination. But it is
the ammunition which does the damage and is the raison d'etre of the small arm.
Guns have no reason to exist except to fire a projectile at a target. In the case of
small arms, the projectile's primary purpose is to incapacitate personnel.

Justification

It is worth questioning the need to keep small arms. After all the proportion of
casualties which they cause is small. This varied from about 15% in the
Mediterranean Theatre of World War II, when armoured forces predominated, to
about 30% in the largely infantry style war waged between the US and Japanese
forces in the Pacific Theatre. However neither percentage tells the full story.
Small arms are vital for many tasks and for many reasons. When all else has
failed it is up to the infantryman to winkle out a really determined defender, and
whatever other aids he has, he would not attempt such a task without effective
small arms. Behind this statement looms the seldom spoken but none the less
well understood factor, that the feel of a small arm in a soldier's hands is a
terrific boost for his morale. He has something with which to hit back, however
ineffectual it may be against some targets. If he did not have this means to fire
back when under pressure he would be far more likely on many occasions to get
his head down or to run. The small arm is also a last ditch in defence. If the
enemy has broken through to rear areas and soldiers of the administrative tail
have to take them on, as has so often happened, they need their small arms.

Factors

Military staffs have now come to appreciate that the ammunition is the all important element in a small arm weapon system, although recent small arms history
tends to suggest that they have had to relearn the lesson the hard way.   The NATO
choice of the 7.62 x 51 mm round in the 1950s could not have been the outcome of
a sound appreciation of the factors solely affecting calibre.

The factors that need to be considered are covered in some detail in Chapter 3.
The basic question is what must the projectile do and at what range? Obtaining
a full answer takes the writer of military requirements into the realms of tactics,
the analysis of data from earlier conflicts, to the consideration of medical statistics, and to an assessment of the capabilities of complementary weapon systems.

Fig 1.1 Infantry with supporting tanks near Le Havre 1944

Tactics

Small arms are employed in all phases of war, all types of conflict, and every
variety of terrain. Whatever his employment on the battlefield - tank crew,
artilleryman, driver, fitter - the soldier has a small arm. The most numerous
corps, the infantry, still counts the small arm as its basic weapon. The weapons'

characteristics must cater for all the likely situations in which soldiers might find themselves. If they work in confined spaces they need small, handy weapons. If their main task requires free hands the weapon must be slung or holstered, but in such a way as to be readily available. Given rain or snow, high temperatures or low, the weapon must function faultlessly. Though their training on small arms may be minimal, the simplicity and handiness of the weapon should be such that it is effective in their hands. Despite all these demands the sheer numbers of the weapons involved dictates that they should not be too expensive. Figure 1.2 serves to highlight the statement that weapons must function satisfactorily under all service conditions.

Fig. 1.2 On NATO's North flank

The enemy's tactics as well as ones own must be taken into account. Heavy sustained volumes of fire were needed to defeat the hordes of Chinese who stormed UN positions in Korea, yet their Communist brethren sneaking through the jungles of Malaya could be dealt with in a short sharp ambush. So the consideration of tactics will affect the requirements for small arms because these weapons are used for so many different tasks in a great variety of environments. The whole spectrum of military engagement, from all out war to internal security, involves using the basic personal weapon of the infantry. The small arm is called upon to beat off attackers, to support the assault on to the objective, and at the other extreme, to kill a fleeting single terrorist in a jungle situation, whether the jungle is tropical or concrete. The extremes of terrain and climate all bring with them demands on the weapon design. In hot desert conditions the weapon must not become so overheated that it cannot be held. If it is humid the sights must not mist up, nor the ammunition exhibit variations in muzzle velocity. In bitter cold a gloved hand must be able to operate the trigger, but the trigger must not be so exposed that it catches on undergrowth to give accidental discharges. The weapon must be light enough so that carrying it on a 20 mile march is not too fatiguing, but it must be sufficiently robust so that the rigours of crawling, jumping and general rough handling do not damage it.

Fig 1.3 A patrol in Malaya

Data Analysis

Analysis of data has become big business since World War II. It was that conflict which gave such impetus to the science of operational research. The extent of world wide conflict over the last 40 years has provided an enormous amount of data. The military staffs who write equipment requirements have access to much information which can help to solve some of their problems. The lessons of previous conflicts can be analysed and advice given on the characteristics needed by future equipment. Thus soldiers can formulate their requirements on a better basis of information and consequently with more confidence that these reflect the true need.

As an example, an analysis of the statistics from conflicts including World War II, Korea and Vietnam show that 95% of all rifle engagements take place at 400 metres or less. As a later chapter on design will show, large calibre ammunition is not needed if the effective range of rifle ammunition is so short. Small calibre ammunition can be just as effective.

The Smaller Calibres

In recent years there has been a widespread move in the small arms world towards smaller calibres away from 7.62 mm or 0.3 inches. The evidence from a variety of sources has clearly indicated that such big rounds were on many occasions unnecessary. Medical statistics, coupled with experimental work on comatose animals and gelatine blocks, have taught an immense amount in recent years about wound ballistics. The damage which a bullet causes varies with its velocity, its calibre and hence mass, its shape and its stability. If the other three parameters can be adjusted to produce an effective bullet, the mass or calibre can be reduced and chosen for other reasons.

The move away from larger calibres in small arms is also partly due to a better appreciation of the inter-relationships of weapon systems and the capability of other warheads to replace small arms fire at intermediate ranges (600-1000m). It is also due to a need to reduce the infantryman's load. Smaller rounds travelling less distance allow the designer to make smaller, lighter weapons. This is a long overdue improvement: the infantryman has been overloaded for years. In certain situations such as protracted jungle patrols his load could be well in excess of 60lbs (27kg). Specialists like radio operators and mortarmen could be carrying even more. Careful field studies have shown that the top weight, over which the infantryman's efficiency rapidly falls off, is about 45lbs (21kg). Furthermore new weapon systems like LAW (the Light Anti-tank Weapon) which could weigh up to 10kg will be coming into service and will have to be carried into action by the infantryman. Such additional loads suggest the need for a reduction elsewhere. Smaller calibres and lighter weapons allow the infantryman to produce a similar weight of small arms fire for less load, or if present load levels are maintained he can markedly increase his fire effectiveness.

This leads to thoughts of the philosophy of small arms fire. The British infantryman has traditionally fired single aimed shots from his personal weapon, and carefully aimed short bursts from his close supporting machine guns. The US infantryman in recent conflicts has adopted what is known as the 'spray the area concept' where neutralising fire relies on sheer volume, however poorly directed. Whilst the US experts are having second thoughts about this concept, the availability of automatic weapons to insurgent groups in many theatres has taught even the UK that winning the fire fight involves firing more rounds at the opposition than he can fire back. Thus automatic fire and large capacity magazines, or belts, become a necessity and to achieve the volume of fire many rounds are needed. Once more this points to lighter, small calibre ammunition. The smaller weapons which are evolving have further advantages in modern warfare.

Most infantry are carried from one battle to another in vehicles, and frequently these are armoured boxes with very restricted space. Getting in and out of them in a hurry is vital, and a short, handy, light weapon is less likely to impede the soldier.

Logistics

Resupply of ammunition in the field is a necessary chore not often practised in peacetime and often made difficult by the fog of war, the shifting nature of forces and the problems of carrying it. For the same number of rounds, smaller calibres will mean less weight and less bulk. When the total reduction for a whole army is assessed the implications for depots and vehicles is considerable.

Another important aspect of logistics with weapons is their ease of maintenance. Servicing and repair takes place at various levels, from the soldier stripping and cleaning his weapon, through the unit armourer carrying out simple repair, to workshops doing major repairs. If the weapon is of a simple robust design the maintenance task is eased. With a modern light weapon the parts themselves should be lighter, and with the use of modern high grade materials the incidence of failure should be less. These trends will reduce the volume of spares which need to be carried and the frequency of repair.

Accuracy

We shall examine in Chapter 5 the various factors which influence the chance of a hit with a small arm. It is a characteristic which illustrates well the balancing act which has to be performed when writing weapon requirements. Accuracy can be greatly increased by using a long barrel, with a long sight base, and firing a high velocity projectile with a flat trajectory. This straight away leads to a heavy design of weapon. Considering the alternative lighter weapons, they are quieter, easier to handle and cause less recoil, all of which tend to improve the shooting abilities of the average soldier. Taken overall, with other factors which favour small calibres, a high degree of accuracy may become desirable rather than essential.

The Family of Small Arms

The generic term 'small arms' embraces various types of weapon. In trying to define them many people include cannons of up to 30 mm. The British Army think of the term embracing weapons up to 12.7 mm or $\frac{1}{2}$ inch. The hand gun is the smallest, being also called pistols, revolvers or automatics depending primarily on their method of operation. Next up the scale are the sub machine guns, some of which today are hardly larger than the biggest hand gun. These very small guns are being designed for use in very close quarter shooting such as is encountered in some anti-terrorist operations. The photograph at Fig. 1.4 opposite compares the size of a conventional pistol with that of a machine pistol.

**Fig 1.4 A Browning automatic pistol
shown for comparison of size with
a Heckler and Koch machine pistol**

Of similar size to many sub machine guns is the carbine, and indeed the terms
become confused with the Americans referring to sub machine carbines. Carbines originated as small rifles which could be carried on horseback, so it is not
surprising that with their mounted cavalry tradition the Americans favour the
name carbine. On the continent the terminology becomes more confused with
machine pistol and assault rifle both tending to describe what the British call
SMGs. The true carbine or small rifle has been overtaken by the modern gene-
ration of smaller individual weapons, and it is this very term which the UK is
now applying to its new rifle - Individual Weapon (IW). Rifles as known for
nearly 100 years will be around for some time to come. They are robust and
effective, accurate to long ranges, and easy to operate and maintain. In certain
situations they are ideal weapons, as for instance the sniping role, and one can
vizualize the Afghan picking off Soviet soldiers from his rocky hideouts as he
picked off British soldiers 100 years ago. Figure 1.5 shows a 7.62 mm SLR
(self loading rifle) and a 4.85 mm prototype Individual Weapon.

**Fig 1.5 A 7.62mm Self Loading Rifle shown
against the Enfield design of Individual
Weapon, seen here in 4.85mm calibre**

Finally there are the machine guns, and as small arms these embrace calibres up to 12.7 mm. Most machine guns at present are in 7.62 mm calibre, and their designation as Light Machine Gun (LMG), Medium Machine Gun (MMG), Heavy Machine Gun (HMG), or Vehicle Mounted Machine Gun (VMMG) depends mostly on their mounting and the extent to which they can sustain fire over long periods. The influence of mounting is shown by the UK designation of General Purpose Machine Gun (GPMG) for the L7A2 version of the FN design. It is used in the light role with a bipod at section level, and at company level in the medium role with a tripod. Figure 1.6 shows the forerunner of the fashion for general purpose machine guns.

**Fig 1.6 Forerunner of the fashion for GPMGs,
the German MG 42**

Now the military requirements staffs have to choose between this wide range of small arms. When small arm tasks are defined, no single one of the many types will fill them all or even a large part of them. There are many good reasons, however, for trying to reduce the variety to only a few. These reasons include logistic considerations such as fewer varieties of ammunition, less spares to carry, interchangeability of parts; training time can be reduced, and the cost of a big buy of one type of weapon is cheaper than many small buys of a variety. The trend is therefore towards the development of a family of weapons which meets the military requirement and involves as few types and calibres as possible. This is well illustrated by the UK suggestion to replace four weapons (the SLR in-service rifle, the GPMG(FN), LMG (Bren) and SMG) with two which are very nearly the same (the Individual Weapon (IW) and the Light Support Weapon (LSW)). The LSW has a heavier barrel and a bipod, and may finally have a somewhat different mechanical action, but in other ways is identical to the IW. Whilst this meets the requirement at section level, the need for sustained fire at longer ranges is not catered for by the IW and LSW. These weapons are expected to be in the 5.56 mm calibre which has recently been adopted by NATO. The heavier NATO calibre of 7.62 mm will be retained for the sustained fire role. The Soviets have developed the well known and widely used family of Kalashnikov weapons. Figure 1.7 shows the AK47 and the RPK LMG which are very similar in design philosophy to the UK IW and LSW.

Fig 1.7 Soviet AK 47 and RPD weapons of the Kalashnikov family

Specialist tasks will still need to be catered for. If the sniper role is retained then a rather more traditional style of rifle will be needed. It will only be required in very small numbers, and may also satisfy a requirement for target rifles for the competition shots in the Army. Some pistols will be kept for personal protection of VIPs and specialists. Despite the virtual uselessness of the pistol in an operational role unless in the hands of an expert, it does have the advantage of not getting in the way and is useful for those who need both hands free to carry out their main task. A few other specialist weapons may be acquired to meet the particular needs of Special Forces. Apart from these specialist weapons the bulk of the nation's small arms will be a small family, often with great commonality.

Commonality is being taken further within NATO by the adoption of not only a standard calibre for the new range of smaller weapons, but a standard specification. Despite there having been a NATO standard calibre of 7.62 mm, interoperability of the ammunition in weapons of that calibre has not been possible because specifications of length, charge weight, materials and method of presentation to the weapon have not, in practice, been standard. Whilst the last element, presentation by magazine or belt, is only being partly addressed with the 5.56 mm NATO calibre, at least all the other specifications are to be the same in all the ammunition manufacturing countries of the NATO allies.

Statement of Requirement

We have looked in outline at some of the ways in which military requirements are arrived at, the factors involved, and some of the thought processes, constraints and balances. The time arrives when the Army must state its requirement. This timing is critical. As much as possible of the analysis mentioned above must be done, and more besides, with the aim of leaving as little as possible to chance. As the old weapons become obsolescent, costly and difficult to maintain, it is imperative to state the requirement for the new in good time to allow for design, for the manufacturing line to be prepared, for trials to take place and for the weapons to be made before the old weapons no longer meet operational needs.

In the UK the Army endorses a document known as the General Staff Requirement (GSR). Other nations have documents with different names but a similar purpose. In USA it is the Requirement for Operational Capability (ROC). The purposes of these documents are to justify the need for the weapon, to persuade the financiers to allocate the money and to give the designer adequate guidance to make his design. For a small arm this will consist largely of a series of characteristics in an order of priority. The designer needs these priorities because so often one characteristic may be enhanced by a design feature, whilst another is actually degraded. Within these guidelines the designer has a relatively free hand. It is only relative in that it is incumbent upon him at all times to ensure that what he is producing satisfies the soldier.

The characteristics are likely to include:

Range   -   a figure for effective fire; this latter expression has to be defined, probably in relation to penetration.

Penetration   -   usually specifying NATO standard targets of mild steel plate at specific ranges.

Wounding effect   -   stated in terms of incapacitation which is explained in more detail in Chapter 3.

Accuracy   -   a difficult term to define and so either stated in relation to the previous weapon system, or in terms of what the average soldier ought to be able to achieve.

Consistency   -   a measure of the weapon's ability to place a succession of shots within a particular area, thus contributing to the accuracy.

Rate of fire   -   not a statement of cyclic rate but the soldier's estimate of what number of rounds he will want to fire in extreme tactical situations with minimum danger of cook-off and little barrel wear.

Reliability   -   expressed in terms of the number of rounds to be fired without failure.

Weight   -   a figure not to be exceeded, stating whether it is with or without particular number of rounds.

Length   -   a desirable maximum.

The performance of the weapon is tested in an extensive and rigorous set of environmental conditions which the soldier must specify. Further requirements are spelled out which are perhaps not so fundamental to the weapon system, but none the less are important, namely: ease of maintenance; ease of training; the sighting arrangements; ammunition types - tracer, training natures, how it should be packaged; attachments to the muzzle such as bayonet, flash hider and grenade discharger and attachments elsewhere such as sling, carrying handle and tools.

Machine Guns

All that has been said so far has tended to refer primarily to the basic weapons, rifle and LMG or IW and LSW. For other small arms and cannon a similar process of analysis is followed to produce a written document from which the designer works.

The basic tactical requirements for machine guns have been touched on. The light version for carrying into the assault or giving final protective fire is usually a bipod mounted weapon, in one of the two NATO standard calibres of 5.56 mm or 7.62 mm. A heavier supporting machine gun is required for neutralisation of troops in the open at longer ranges or taking on soft skinned vehicles. This will in all likelihood also be of 7.62 mm calibre, but the type of mounting will confer on it a steadier base so as to produce the required accuracy. Spare

barrels changed at predetermined intervals of time or after a set number of
rounds will relieve heating problems.

Vehicle mounted machine guns (VMMG) are coming into line on calibre. Old
faithfuls like the Browning .50 inch are being replaced by weapons giving at least
commonality of ammunition. Most British vehicles have for instance been mounting a version of the FN design, basically similar to GPMG. These guns vary in
their ability to be used in the dismounted role (L37) or not (L8).

Fig 1.8 The L8 version of the GPMG which is
not dismountable

Refinements being looked for in VMMGs include: an even greater standard of
reliability than ground mounted weapons, since clearing misfires in vehicle tur-
rets can be awkward; a considerable reduction in the toxic fumes caused in tur-
rets; and improvements in mounting to facilitate such tasks as changing barrels.
Traditional gas operated weapons, purpose designed for armoured fighting
vehicles (AFVs) can go a long way towards fulfilling these criteria, but not all of
them. The tendency is to look to revolutionary designs such as are mounted in
many modern aircraft and which achieve what is needed because they are externally powered. Examples are the Mini gun, the Vulcan gun, the Hughes Chain
gun and the Dover Devil. The photograph at Fig. 1.9 shows the GEC 7.62 mm
Mini gun.

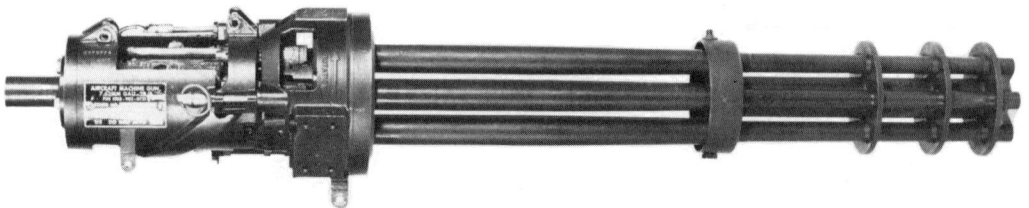

Fig 1.9 The GEC 7.62mm Minigun

Gas is not needed for operation so it can be channelled out of the vehicle, as can the empty cases. Rates of fire and reliability are said to be very good. The working parts are small so that less space is taken up in the turret, and they can be hinged away to allow the barrel to be changed. There are drawbacks, such as the reliance on a power take-off from the vehicle (although hand cranking may be possible with some) and the ability to dismount and use the weapon in the ground role is virtually lost.

Heavy machine guns (HMG) in bigger calibres than 7.62 mm used to be employed against lightly armoured targets. Light armour, in the sense of being easy to defeat, is a thing of the past, and so too is the HMG. The replacement of such weapons as the HMG shown below in Fig. 1.10 is a cannon.

Fig 1.10 A 12.7mm HMG Degtyarev

Requirements for Cannon

The ever more widespread use of infantry combat vehicles (ICV) is leading to-wards a traffic congestion on the future battlefield not previously experienced. The use of expensive tank main armament and guided missiles, in short enough supply already, to take out ICVs may not be cost effective. The requirement therefore arises for the ICVs to have their own means of destroying their op-posite numbers. As armours improve the assured destruction of such vehicles as the Soviet BMP can only be accomplished with cannon, and fairly large calibre cannon if frontal glacis plates are to be defeated. Cannon generally are in 20, 25, 30 and 35 mm calibres, though it is even possible that NATO may pick up another calibre of 27 mm due to interest in the Mauser gun mounted in Tornado which is being adapted for vehicular use.

The argument in NATO for adoption of a common cannon calibre is being only partly driven by military considerations. Quite large commercial interests are involved as cannon ammunition is far more profitable even than small arm rounds. Whilst the biggest calibre finds favour with those who see lethality as the dominant characteristic in the weapon, the size of the gun and the ammunition storage problem becomes difficult for those nations like the UK, who demand capacity as the foremost priority in their ICV. A compromise at 25 mm or 30 mm, the next stage down, is a likely choice therefore. It has the added advantage that those concerned with causing damage to aircraft also tend to favour this calibre. NATO countries making air defence cannon are obviously going for lethality with 35 mm (FRG) or 40 mm (US) calibres. The USSR on the contrary has invested heavily in 23 mm, exchanging quantity for calibre.

Variation in Requirements

It is salutory to conclude this chapter with a recognition that requirements in different countries are not in accord with one another, even among allies. This illustrates not so much a failure on the part of those who write the requirements as an interesting study of the genesis of the requirement. If an ally, or even an enemy, is at variance on a small arm requirement it is wise to ask why. Have we got it wrong or have they? No one can answer that for certain until the conflict for which they were designed takes place - hopefully never. Quite often it may be that neither necessarily has got it wrong. With the scales frequently finely balanced in argument for or against a characteristic it is easy to come down on different sides. Small influences of historical experience, precedent, designer's quirks, or the hobby horse of a senior officer may be enough to swing the balance. Of these, recently remembered military history can be the most vital. The Germans for instance are said to be most reluctant to move to small calibres and are sticking with 7.62 mm for all section weapons because of their battles of attrition with the Red Army in World War II. The cynic may say the reason is more commercial. FRG is the closest of all the NATO partners to breaking the technological barrier leading to the introduction of a caseless ammunition. Why switch to 5.56 mm when in 10 years they could lead the world with a revolutionary small arm?

Tactical philosophy may make nations choose differently. The Soviet and Chinese armies can still count on the use of infantry in large numbers and hence en masse. The less numerous Western European infantry are far more inclined to use fire and manouvre tactics, and the infantry weapons to support these tactics are inevitably going to be different.

When all is said and done on requirements, national economic considerations often hold sway when the decisions are taken. Small arms, as was pointed out earlier, are such a fundamental requirement that no country will wish to give up a capability to make them. If they have that capability the financial and employment benefits are also very important. Countries with quite small armies may have fairly large armament industries, Sweden and Belgium being good examples in the small arms field. The influence their requirements staffs can have is as nothing compared with the industrial lobby of FFV and FN. Industrial concerns take great trouble with market research, and the views of world wide potential customers are therefore reflected in their designs. Compromise is inevitable, but whilst strong and sometimes extreme views may be voiced about the demerits of a particular weapon, few modern ones are anything but excellent. They will not sell if they are not.

One small arms manufacturer at least has been heard to say that he will give away his weapons if the customer will buy his ammunition. So we have come full cycle from the soldier recognising the importance of the ammunition to the manufacturer knowing that that is where the profits lie. The small arm is the means of delivery.

SELF TEST QUESTIONS

QUESTION 1   As small arms are responsible for very few casualties in war,
              what reasons are there for their retention?

              Answer   ....

                       ....

                       ....

QUESTION 2   What foremost lesson has data analysis provided to guide small
              arm requirements?

              Answer   ....

QUESTION 3   List some factors which favour the use of smaller calibres.

              Answer   ....

                       ....

                       ....

QUESTION 4   List as many characteristics for a small arm requirement as
              you can.

              Answer   ....

                       ....

                       ....

                       ....

                       ....

                       ....

QUESTION 5   What are the distinctions between LMG and MMG?

              Answer   ....

                       ....

QUESTION 6   What are the drawbacks of mounting LMGs designed for the
              ground role in vehicle turrets?

              Answer   ....

                       ....

QUESTION 7   What environmental hazards must a designer cater for in a small
                arm?

                Answer   ....

                           ....

QUESTION 8   How have medical statistics and experiments influenced the
                small arm requirement staffs and designers?

                Answer   ....

                           ....

QUESTION 9   Why are HMGs obsolete?

                Answer   ....

                           ....

QUESTION 10   Why might national requirements differ between countries for a
                small arm which is to fulfil a similar role?

                Answer   ....

                           ....

ANSWERS ON PAGE 179

SAC - C

2.

Basic Scientific Principles

INTRODUCTION

When designing a weapon the first consideration is to look at the likely target and decide what should be done to it. When this has been resolved by the soldier the weapon designer determines what sort of projectile is required to produce the desired damage at the target. In some instances the most effective projectile may be the most difficult to deliver. Thus the design of the selected projectile may have to be modified to deliver it accurately to the required range from a suitable projector. This narrower set of possible projectiles can then be examined with a view to producing the design of the most effective projector. The interaction between the various limiting effects of the laws of physics on first the projectile, second the propellant and finally the projector are such that, in practice, they cannot be considered separately. However, in this chapter the subject will be covered in stages in that logical sequence working back from the desired effect at the target.

TARGET EFFECT

The normal target for small arms ammunition is the human body. The desired effect on this target is a lethal or incapacitating wound. The lethality of a wound depends on its location and on the nature of the wound itself. The soldier asks for ammunition which will be rapidly effective no matter where the target is hit. One of the tasks of the ammunition designer is to study the wounds produced by various types of projectiles. The effects of different wounds on the functioning of the body are examined with the co-operation of those knowledgeable in anatomy and physiology. The designer uses the facts gained from this research to produce a warhead which is capable of giving an incapacitating wound and at the same time meets the design requirements of shape, size and weight demanded by the weapon designer.

Attempts to formulate target effect have led to the use of the terms 'probability of incapacitation' and 'transfer of energy'. The rest of this section will examine what is meant by these terms.

19

Incapacitation

Any wound causing loss of consciousness, blindness or paralysis must be incapacitating because the wounded man can neither continue his task nor retaliate.
Other wounds may be incapacitating depending on the location of the wound in
relation to the task. Motivation to carry out his task may enable the wounded man
to overcome what would seem to be an incapacitating wound. Thus the task and
the soldier's motivation need to be considered. This is best illustrated by the
following example: a wound in the foot is more likely to incapacitate an assaulting
infantryman than a static defender. If the attacker were to receive a flesh wound
in the arm he could well continue the assault whereas a similarly wounded defender might find it difficult to continue firing. If the defender does not continue
firing the attacker could reach his position to kill or capture him. To avoid such
a fate the defender would be motivated to fight on. The attacker who is wounded
in the foot would however not similarly be motivated, since by stopping his assault
he would not be so directly contributing to his own death or capture. Thus if the
subjective factors of motivation were ignored by the scientist in any theory of
wound ballistics it would be right to question the value of its results.

The time taken for a wound to become incapacitating is an important consideration.
Only certain parts of the human body are vulnerable to instant incapacitation.
These are confined to parts of the head, the heart and the spinal column above the
third vertebra. As these areas cover about 15% of the body there is at best only
a 15% chance of a random hit on a standing man causing incapacitation within a
few seconds. Therefore there is an 85% chance that, for a random hit, there will
be an interval between the strike of the bullet and incapacitation. The time delay
will depend on the severity of the wound, its location and the motivation of the
wounded man. Of course the soldier on the battlefield would like the result of his
fire to be instantaneous. To guarantee this for a random hit the size and weight
of the ammunition would have to be so great that it would be impossible to design
a small arms weapon to fire it. As a compromise 30 seconds is the shortest time
interval used in assessments of the battlefield wounding performance of warheads.

There are 14 official battlefield wounding criteria of which the Defence 30 Second
and the Assault 30 Second are most often quoted in warhead effectiveness studies.
The Defence 30 Second Criterion is the most difficult to achieve as much of the
target is considered to be undercover, the exposed portion largely protected by a
steel helmet and the soldier strongly motivated to continue firing to prevent his
position being overrun by the enemy.

Energy Transfer

The effects on the human body when struck by a projectile have been closely
studied. It has been found that tissues rupture because the body cannot readily
absorb energy. The amount of damage to the body depends on the level of energy
transferred from the projectile and on the rate of transfer. Energy can be expressed in terms of mass multiplied by the square of its velocity (mv^2). The
transfer of energy is more easily assessed mathematically by considering the
projectile rather than the human target. Energy is transferred to the target when

the projectile loses velocity as a result of striking the target.   The amount of
energy transferred can be calculated by using the following formula:

$$E = \tfrac{1}{2} m (v_1{}^2 - v_2{}^2)$$

Where: E = Energy transferred,

m = Projectile mass

v_1 = Impact velocity

v_2 = Exit velocity

Low velocity blunt projectiles such as flying debris and baton rounds may cause
bruising, haemorrhage and bone fractures.   Similar effects can be caused by
bullets which are stopped by body armour or other items of equipment.   For a
projectile the size of a bullet to cause effective damage it must penetrate the
body.   When bullets enter the body at less than 400 m/sec the damage is restricted to the bullet track.   At higher velocities the impact of the bullet creates a
shock wave which enlarges the track cavity produced by the bullet by up to thirty
times.

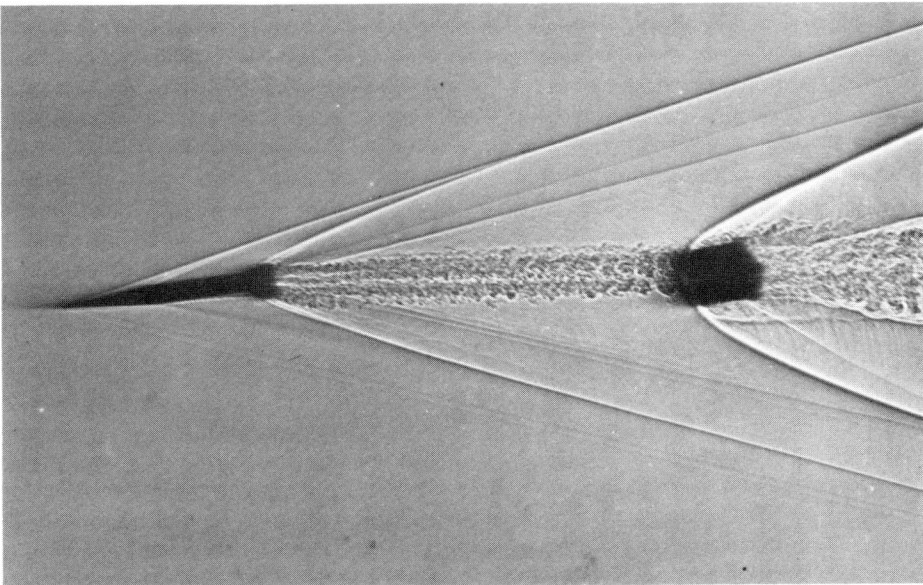

Fig 2.1 Shock Wave Produced by High Velocity Projectiles

The photograph at Fig. 2.1 shows a shock wave in the air.   When a comparable
shock wave spreads through the body of the target the area of damage is greatly
extended and the chance of incapacitation increased.   This effect has been termed
'explosive wounding' and has been a feature of most modern high velocity bullets.
The table overleaf shows the muzzle velocities of bullets fired from some modern
weapons.

TABLE 2.1 Muzzle Velocities of Bullets Fired from Various
Modern Weapons

Bullet mm	Weapon	Muzzle Velocity m/sec
7.7x56R (.303)	UK No 4 Bolt Action Rifle	748
7.62x51	UK SLR L1A1	838
7.62x39	Soviet AKM Assault Rifle	715
5.56x45	US Armalite Rifle	1000
9x19	UK L2A3 Sterling Sub Machine Gun	390

In attempting to analyse wounding effects, work has centred on firing projectiles
of differing shapes, weights and velocities into materials simulating human flesh
and recording the results. The probable paths of the bullets through the body
have been traced by medical experts and the damaging effects assessed. It has
been possible to establish a relationship between the effectiveness of wounding
and a projectile's mass, velocity and shape. The term $mv^3/2$, a refinement of
mv^2 has been produced to explain the results of these experiments. However,
for simplicity the slightly larger term mv^2 will be used throughout this book.

DESIGN OF THE PROJECTILE

To transfer the maximum amount of energy a projectile should have high impact
energy and should not pass out of the target. High impact energy can be achieved
by high muzzle velocity and a projectile whose physical properties allow it to
lose as little of that velocity as possible during flight. The projectile will stop
in the body or its retardation will be increased under the following conditions.
Firstly, if the projectile breaks up within the target, the pieces, having smaller
mass individually, have less penetrating power than the original projectile.
Secondly, if the projectile deforms on impact into a less streamlined shape.
Thirdly, if the projectile becomes unstable when entering a denser medium than
air.

The target is often protected by some artificial or natural material which has to
be penetrated. A bullet stable enough to penetrate cover will also penetrate and
pass out of the target if it strikes it with sufficient velocity. High velocity bullets
transfer only a small proportion of their energy to the target before exiting.
However the level of energy transferred is still high enough to cause effective
wounding.

THE WEAPON

When a round is fired inside the chamber of a small arm the propellant burns and produces a gas which expands to some 14000 times the volume of the original propellant. This gas is burnt and the flame temperature can be as high as $2000^{\circ}C$. As shown in Fig. 2.2, the pressure produced drives the bullet out of the case and into the lead.

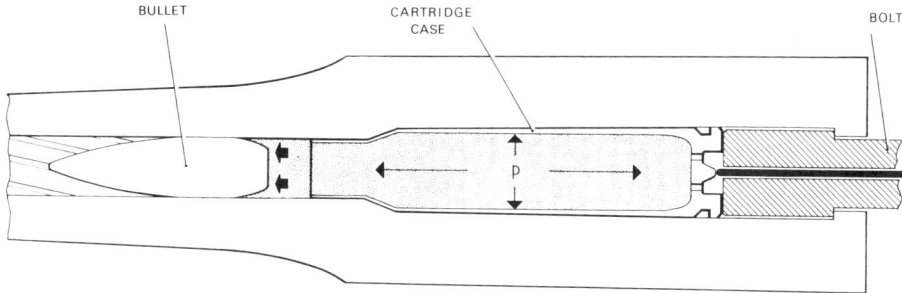

Fig 2.2 Effect of Pressure on the Bullet

The lead aligns the bullet to meet the rifling and holds the bullet in contact with the initial part of the rifling until sufficient pressure has built up to force the bullet down the bore. The diagram at Fig. 2.3 shows the relationship between pressure inside the barrel, the velocity of the bullet and the time interval between cap strike and the emergence of the bullet from the rifle.

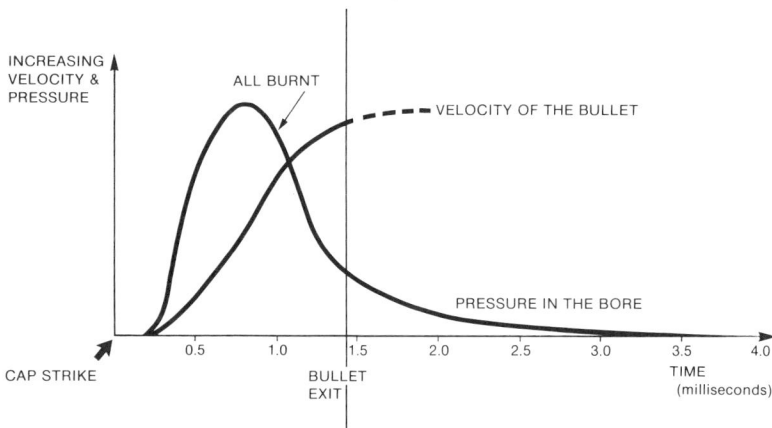

Fig 2.3 Relationship between Pressure, Velocity and Time

As the projectile moves down the barrel the chamber volume beind the bullet in-creases. Shortly before "all burnt" the rate of increase of chamber volume is greater than the increase in gas production and so the pressure starts to fall. When the bullet leaves the muzzle of the UK L1A1 SLR barrel the pressure is approximately 90 MN/m^2 (Mega Newtons per square metre) and this pressure then falls to atmospheric about 2.5 to 3 milliseconds later. Figure 2.4 shows the L1A1 rifle pressure/time relationship in more detail and indicates the part of the weapon subjected to maximum pressure.

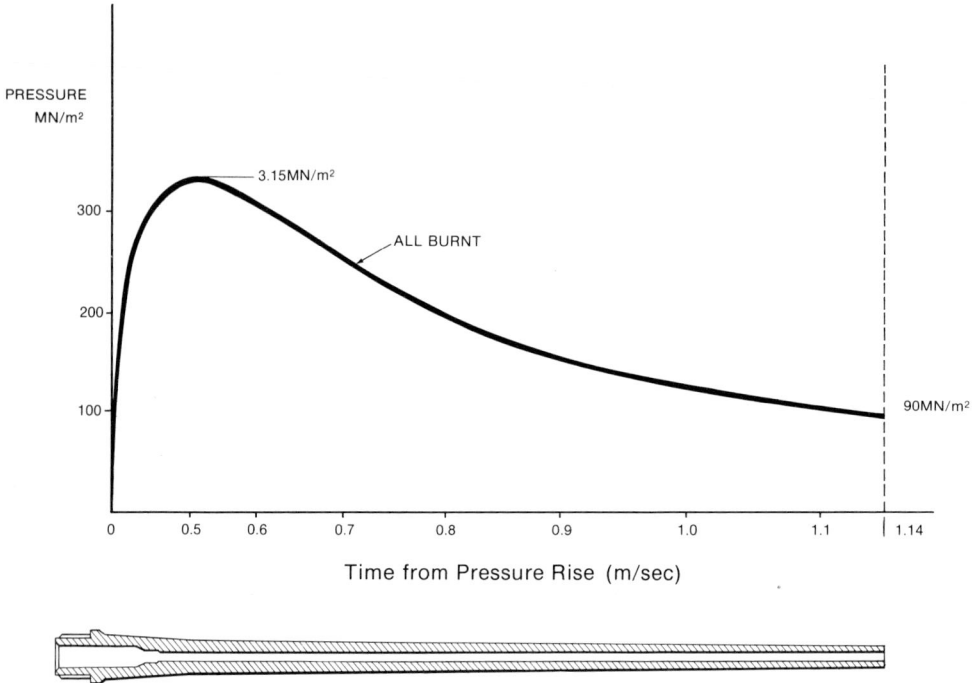

PRESSURE MN/m²

3.15MN/m²

ALL BURNT

300

200

100

90MN/m²

0   0.5   0.6   0.7   0.8   0.9   1.0   1.1   1.14

Time from Pressure Rise (m/sec)

Fig 2.4 Pressure/Time Curve in the UK L1A1 Barrel

Recoil

When the propellant burns it exerts pressure on the chamber in every direction, as shown in Fig. 2.2. This pressure forces the bullet along the bore and drives the breech block backwards. If the breech block is secured to the barrel and body, then the whole weapon is pushed back causing what is known as recoil. The forward momentum of the bullet is expressed as its mass (m) multiplied by its velocity (v). The backward momentum of the rifle can correspondingly be written as MV. The momentum given to each of these must be the same as they experience the same force over the same time and as both start from rest then both must have the same momentum when the bullet reaches the muzzle. Therefore mv = MV. It is important to note that a weapon's momentum depends on the momentum contained in the round. A round which gives increased momentum to the bullet will similarly increase the weapon's momentum. If the mass of the propellant gases is ignored, the free recoil velocity of the weapon can be expressed in a simplified way as $V = \frac{mv}{M}$. This means that for a given muzzle velocity and a given weight of bullet, the velocity with which the weapon moves back is governed by its mass. The heavier the gun the slower it moves back after being fired. The muzzle velocity and the weight of the bullet are of course determined by the tactical requirements for the weapon. The ratio between the weight of the weapon and its velocity are influenced by the amount of energy a soldier can be expected to absorb in his shoulder. The kinetic energy of the weapon is $\frac{1}{2}MV^2$ and in practice the upper recoil limit has tended to be around 15 Joules. Given the bullet mass and its muzzle velocity the designer can use the following formulae to determine weapon mass.

If the recoil R is expressed as: $\quad R = \frac{1}{2}\,MV^2$

and the velocity: $\qquad\qquad\qquad V = \dfrac{mv}{M}$

Then by substitution $\qquad\qquad R = \frac{1}{2}\,M\left(\dfrac{mv}{M}\right)^2$

$$= \tfrac{1}{2}\,\dfrac{(mv)^2}{M}$$

or $\qquad\qquad\qquad\qquad M = \tfrac{1}{2}\,\dfrac{(mv)^2}{R}$

Therefore all weapons which weigh the same and fire the same round will produce an identical recoil. It is also true to say that if the weight of the weapon is excessively reduced the increased recoil will make it difficult and unpleasant to handle.

Barrels

The longer the barrel the greater the velocity because the bore pressure acts for longer. However, as shown in Fig. 2.3 any increase in barrel length after the point of 'all burnt' gives only a proportionately small increase in muzzle velocity. The advantage of this increase is probably offset by the weapon becoming too long to be easily managed, particularly when being moved in and out of vehicles. On the other hand the barrel could be shortened with a corresponding small loss of muzzle velocity. This would result in a slight reduction in weapon weight but have the disadvantage of increasing recoil. Barrel length is generally governed by balancing these factors which are discussed in more detail in Chapter 3.

Barrel walls must be strong enough to withstand the pressure of the expanding gases and thick enough to minimise overheating problems. They must also be strong enough to stand up to rough handling and, in the case of some rifles, projecting grenades and bayonet fighting.

There are two main stresses acting on the barrel material when the bullet is fired. These are shown in Fig. 2.5. Radial compressive stress acts outwards on the walls of the barrel. This stress is at its maximum at the inner surface and decreases towards the outside. Hoop tension stress also occurs because the barrel material is stretched circumferentially. This stretching effect diminishes towards the outer barrel surface. Hoop tension stress is always greater than radial stress and, in theory, is the limiting design factor in barrel strengths.

It is apparent that the material nearest the inside of the barrel is the most stressed. It is possible to distribute these stresses more evenly by prestressing.
However this process, called autofrettage, is not deliberately included in the
manufacture of small arms as barrel walls thick enough to be 'soldier proof' and
to overcome heating problems are also thick enough to withstand firing stress.
The subject of autofrettage is covered in more detail in Chapter 4 of Volume II,
which is the book on Artillery.

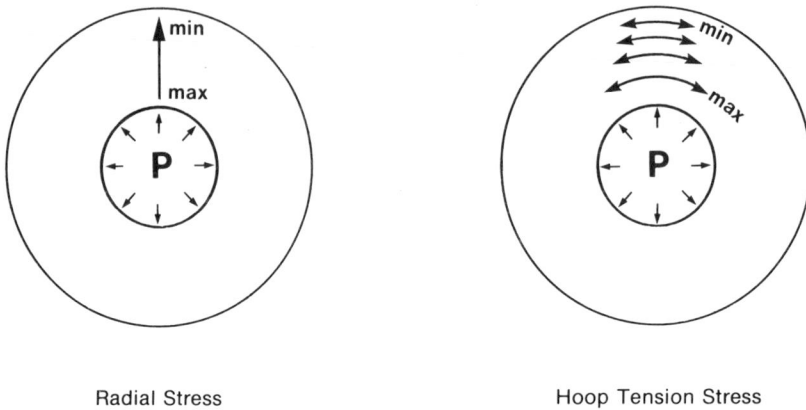

Radial Stress   Hoop Tension Stress

Fig.2.5 Barrel Stresses

Table 2.2 shows the overall barrel diameters just forward of the chamber for a
variety of weapons. It demonstrates how barrel walls have been thickened in
order to solve the heating problems of rapid fire. Those weapons with the lowest
ratio of bore to barrel diameter (relatively thin walls) have no automatic fire
facility.

TABLE 2.2 Barrel Diameters of Various Weapons

Weapon	Nominal Weapon Calibre (mm)	Outer Barrel Diameter (mm)	Ratio
Austrian Steyr	5.56	18.5	1:3.3
US Stoner LMG	5.56	18	1:3.2
UK L1A1 SLR	7.62	19	1:2.5
UK L7A1 GPMG	7.62	30	1:3.9
UK No 4 Rifle	7.7	20	1:2.6
Vickers MMG	7.7	23	1:3.0

Bullets are made stable in flight by the spin imparted by the rifling inside the
barrel. The twist of the rifling is the length required for one complete turn of

the bullet. Normally the twist is one in thirty calibres but this can be reduced for high velocity or short bullets. If the twist is very tight the bullet jacket can become overstressed. Lessening the twist reduces the potential stabilising effect. It is immaterial whether the direction of the rifling is to the left or, as in the case of most weapons, to the right. The diagram at Fig. 2.6 shows the main features of a rifled barrel. The ridges in the metal are called lands and the space between them grooves. The internal dimensions of the UK L1A1 SLR are included in the diagram to show that the nominal 7.62 mm calibre does not refer to any bore dimensions.

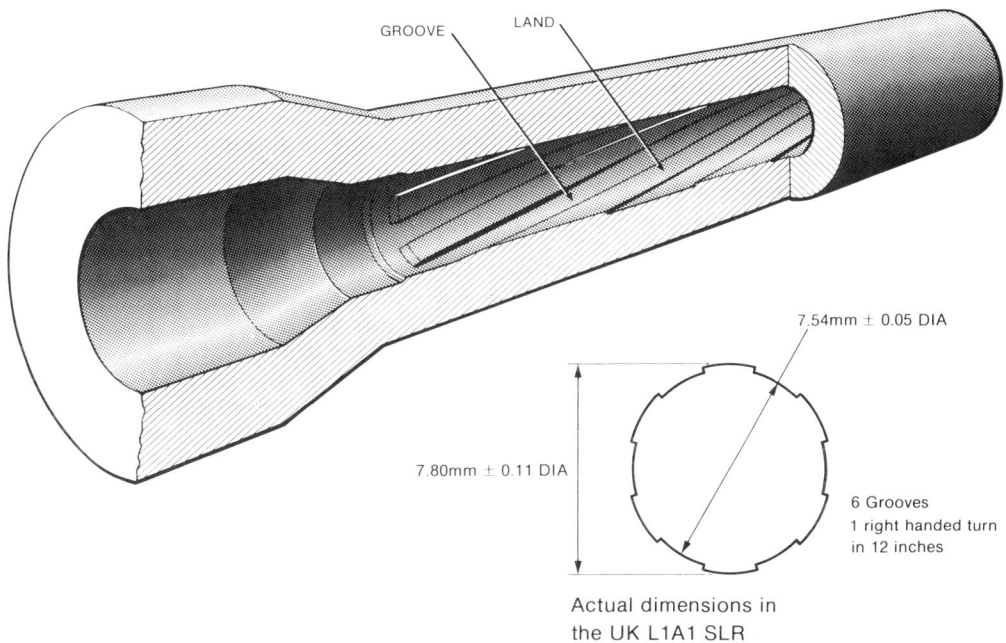

GROOVE   LAND

7.54mm ± 0.05 DIA

7.80mm ± 0.11 DIA

6 Grooves
1 right handed turn
in 12 inches

Actual dimensions in
the UK L1A1 SLR

Fig 2.6 Sectional diagram of rifling in a barrel

Wear

The barrel must be strong enough to resist the abrasive wear caused by the bullet moving up the bore and the erosive action of hot, high pressure gas. All military weapons have steel barrels in order to provide the necessary strength at an economic price.

Abrasive wear occurs throughout the length of the bore but is most apparent at the commencement of rifling. When a bullet passes down a rifled barrel the lands of the rifling engrave the bullet jacket. This causes high pressures in the order of $450MN/m^2$ between the faces of the lands and the bullet. Once engraving is complete pressure is reduced. Barrel steels are strong enough to resist deformation until a temperature of approximately 500^oC is reached when the hardness of steel can be reduced to below that of the metal comprising the bullet jacket. This causes extensive damage to the lands. As described in greater detail in Chapter 5, the bullet base expands (sets up) to fill the grooves in the rifling. The whole of the bore therefore comes in contact with the bullet so that wear occurs over the whole surface of the bore. If gas were to escape through any narrow aperture between the bullet and the bore it would produce intense local heating which might even melt the bore surface. This effect is called gas wash erosion. A little gas is likely to escape before the bullet is fully engaged in the rifling so that it is the commencement of rifling which is subject to this type of erosion. The photograph at Fig. 2.7 shows the wear that has occurred to the bore of a machine gun barrel. The damage is quite pronounced at the commencement of rifling.

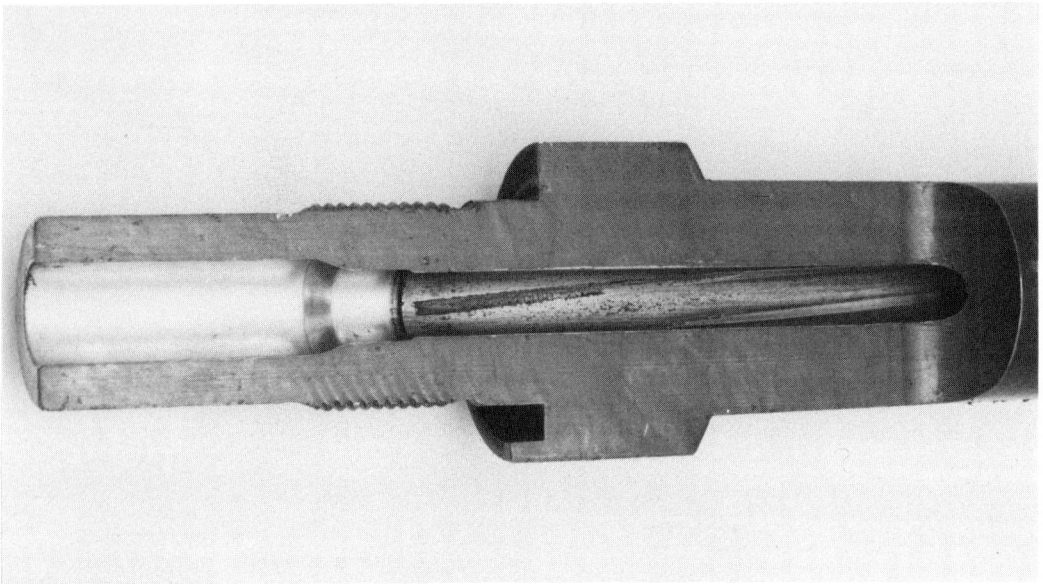

Fig 2.7 Wear inside a machine gun barrel

The high pressure, hot gases can also transform the surface metal of the bore into a brittle layer. Thermal stresses in the metal crack this layer which is then rubbed away by the passage of the next bullet. The time taken for the formation of this brittle layer is longer than the interval between shots when a gun is firing on automatic. Thus this type of damage occurs when single shots are fired at intervals.

Heating

Barrel heating can be a major problem in the design of automatic weapons. When a weapon is fired, escaping hot gases transfer their heat to the barrel. The chamber will eventually become hot through conduction from the barrel but is

initially shielded from the heat source by the cartridge case. Therefore premature ignition of propellant, often called 'cook off', only occurs when heat has had
time to spread to the chamber from the barrel. If caseless ammunition were
used the hot gases would be in direct contact with the chamber walls which would
therefore heat more quickly. This would lead to a greater chance of 'cook off'.

Distribution of Energy

Only about one quarter of the energy available in the propellant is used to propel
the bullet down the barrel. The disposal of the remaining energy depends on a
large number of inter-related factors such as weapon weight, barrel mass and
barrel length. An approximate guide to the distribution of energy is shown below.

Provision of kinetic energy for the bullet	20 - 30%
Heat to the barrel	30%
Muzzle blast	Up to 40%

Not shown in the above are the very small amounts of energy which are used to
provide rotation of the bullet or overcome friction. It is of note that only about
0.1% of the available energy appears as recoil.

SUMMARY

This chapter has covered, in outline, the required target effects of small arms
and the basic scientific factors governing their design. Since there is an 85%
chance of a random hit proving non-fatal the designer aims at causing 'incapacitation' or, in other words, rendering the target incapable of carrying out his
task. The chance of a wounded soldier remaining active for a stated period is
calculated by applying medical knowledge to a statistical system. The most
damaging projectile is one whose effect spreads widely beyond its track through
the body. High velocity projectiles which create a shock wave cause this type of
'explosive wounding'. Bullets which tumble on impact also cause extensive
damage but are unable to penetrate cover and it is this which is seen to be the
dominant factor in ammunition design. The amount of damage caused to the
target depends on the amount of energy transferred from the projectile. The
momentum of the projectile influences not only its effective range but also the
weapon weight and the recoil felt by the firer.

SELF TEST QUESTIONS

QUESTION 1   What physical properties influence the wounding efficiency of bullets?

Answer   . . . .

. . . .

QUESTION 2   What is meant by the term 'incapacitation'?

Answer   . . . .

. . . .

QUESTION 3   What factors influence the effects of any wound?

Answer   . . . .

. . . .

QUESTION 4   What do you understand by the term 'explosive wounding'?

Answer   . . . .

. . . .

QUESTION 5   What are the results of shortening the barrel?

Answer   . . . .

. . . .

QUESTION 6   What decides the minimum acceptable thickness of a barrel?

Answer   . . . .

. . . .

QUESTION 7   What are the effects of altering the twist of rifling in a barrel?

Answer   . . . .

. . . .

QUESTION 8   At what part of the barrel does wear become most noticable and why?

Answer   . . . .

. . . .

QUESTION 9   What is the meaning of the term 'cook off'?

              Answer   ....

                        ....

QUESTION 10 Why is caseless ammunition likely to cause a problem?

              Answer   ....

                        ....

ANSWERS ON PAGE 180

3.

Factors Affecting Choice of Calibre

BACKGROUND

The distance over which a bullet can be fired depends on the amount of propellant used and on the carrying power of the bullet. The amount of propellant depends on the chamber volume available, the power of the propellant and on the quality of the material used to make the weapon. As a broad generalisation the carrying power of bullets made of identical material is proportional to their diameter. With these basic facts in mind, it is of interest to trace the development of small arms since the design of modern weapons is inevitably influenced by historical factors.

Early Weapons

The earliest hand guns were made of poor material and were incapable of withstanding a substantial gas pressure. In order to obtain sufficient muzzle velocity to be lethal at adequate battle ranges it was necessary to have a large diameter bullet. The musket, for example, was originally 10 bore (.79 ins) and was later reduced to 11 bore (.753 ins) with the introduction of Brown Bess in 1700 which fired a ball weighing about 32 gms with about 4.5 gms of propellant. The propellant was black powder which produced so much smoke that the infantry had to use volley fire. The muzzle velocity was so low and so variable that engagements over 80m were unusual. The Brown Bess went out of first line service in 1824.

With the introduction of rifling, more powerful propellants and better gun steels, the calibres of weapons were gradually reduced and at the same time engagement ranges increased. The Baker Rifle (1800) which used the same mass of propellant as the Brown Bess fired a .612 calibre lead ball and was sighted for 100 and 200 yards. This shows the improvement that was obtained largely by introducing a satisfactory system of rifling. However, care should be taken in equating performance with sight settings as these are not always a true indication of weapon performance. The Enfield P/53 Rifled musket of 1853 had a bore of .577 inches and was sighted up to 800 yards. The P/53 when converted to breech loading and

renamed the Snider Rifle (1864) fired a 31 gms bullet and used about 4.5 gms of
improved gunpowder. The Martini Henry Rifle (1871) had a calibre of .45 inches
and was sighted up to 1200 yards. It also used a 31 gms bullet but the amount of
propellant was increased to 5.5 gms.

Recent Weapons

By the end of the 19th century it was possible to fire bullets over considerable
distances. The requirement was for the Infantry to be able to engage targets at
long ranges. The role of the machine gun as the primary long range small arm
had not been fully appreciated and riflemen were expected to engage targets out to
2000m. The Lee Metford Mk 1 Rifle, for example, was sighted to 2800 yards
although this was a little optimistic.

The Lee Metford ammunition originally consisted of a 14 gms lead alloy bullet of
.303 inch calibre plus 4.5 gms of gunpowder. The gunpowder was shortly replaced by cordite producing the first smokeless propellant. The long Lee Enfield
rifles used in the Boer War proved cumbersome and in 1902 the barrel was shortened and the RSMLE (Rifle, short, magazine, Lee Enfield) was introduced as
the No 1 Rifle. It was soon realised that machine guns were more successful than
rifles at long range. However it was felt necessary for the Infantry to have common ammunition for both rifle and machine gun. The .303 bullet which weighed
11.3 gms was retained until replaced by the current NATO 7.62 mm (.300 ins)
round whose bullet weight dropped to 9.35 gms.

ANALYSIS LEADING TO CHOICE OF ROUND

The choice of round depends on both tactical and technical considerations.
Initially tactical demands give an indication of the required performance of a bullet. The application of technical factors results in a variety of options to meet
this tactical performance. The ensuing evaluation makes it possible to specify
the round and gives the outline design of the weapon to fire it.

At Fig. 3.1 is a schematic layout of the technical and tactical factors involved in
the process of choosing a round. For the sake of clarity this diagram over-
simplifies the process in some instances.

The designer must be informed about the target and the required range and
accuracy of the weapon. He then calculates the effect of bullet, chamber and
barrel design on the muzzle velocity of bullets possessing the necessary impact
energy at the required range. Often some compromise with the initial military
requirement is involved. The calibre of the round is virtually fixed at this stage.

Technical Factors   ENEMY SOLDIER   User Requirement

Wound Ballistics   Protection/cover in front
Assessments   of soldier
                                                                     Speed of Incapacition
                                                                     Soldier's Task

                                   STRIKING ENERGY
                                   OF BULLET

Bullet Mass   Range to Target
Bullet calibre   HIT   Acceptable wind deflection
Bullet shape   CHANCE   Acceptable recoil in shoulder
Weapon expansion ratio   Acceptable vertex height
                                                                     of trajectory

                                   MUZZLE VELOCITY
                                   OF BULLET

                                                            Recoil
Bullet calibre and design
Bullet mass   WEAPON MASS
Max chamber pressure

                                   CHARGE MASS

Loading Density
Bullet diameter   CASE DESIGN
Bullet length
inside case

                                   DETAILS OF
                                   COMPLETE ROUND

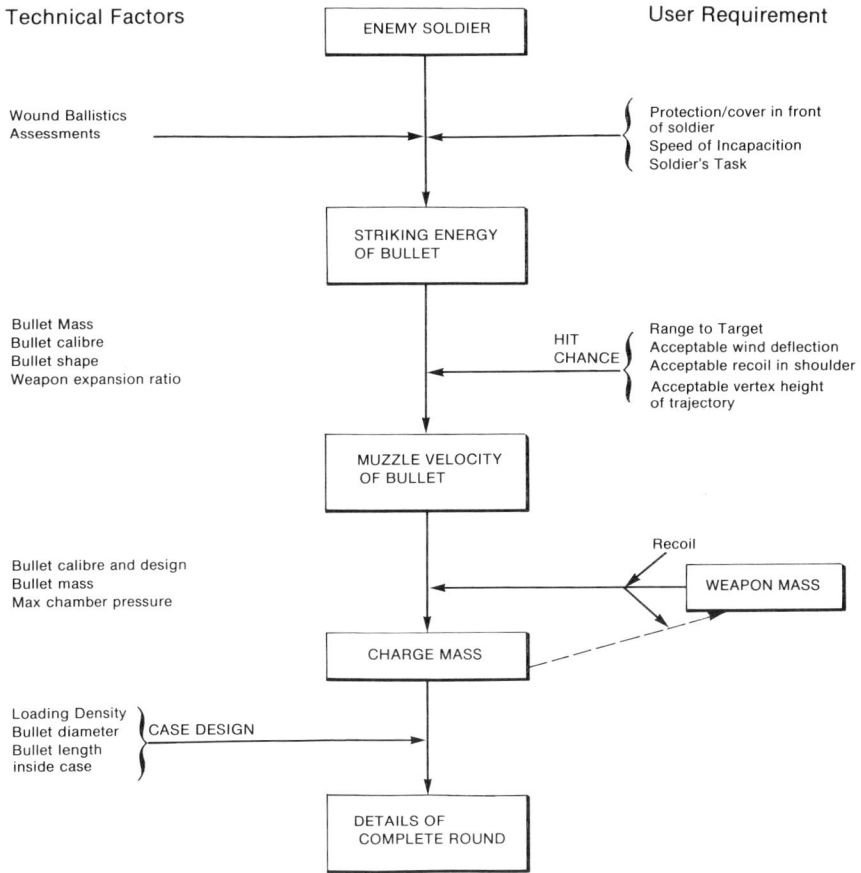

Fig 3.1 Choosing the Round

Weapons of the same nominal calibre are not necessarily capable of firing the
same round. For example, both the NATO and Warsaw Pact Forces use 7.62 mm
weapons but the ammunition is not interchangeable as can be clearly seen in the
photograph at Fig. 3.2 overleaf.

SOVIET   SOVIET   NATO

M1908   M1943   L 2 A 2

7,62x54R mm   7,62x39mm   7,62x51mm

Fig 3.2 NATO and Soviet 7.62mm rounds

To avoid ambiguity it is better to state both the nominal calibre and the case length of the round which the weapon chambers. NATO Forces use weapons chambering a rimless 7.62 x 51 mm round. The Soviet MMG fires a rimmed (R) round 7.62 x 54 mm R whilst the AKM Assault Rifle fires the shorter 7.62 x 39 mm round. When a weapon is able to chamber a choice of rounds then its type number should also be mentioned. For example, one of the rounds fired by the US AR15 Armalite Rifle is described as the 5.56 x 45 mm M193. The next four sections of this chapter will go into certain aspects of the design process in more detail, starting with the effects required at the target.

LETHALITY

Lethality is a term used to describe the efficiency of a warhead. To be effective the bullet must transfer some of its energy to the target. An energy level of 80 Joules is accepted as the minimum required to incapacitate a target within 30 seconds of wounding. As mentioned in Chapter 2, most small arms assessments use the Defence 30 Second Criterion.

Figure 3.3 shows the energy which a variety of bullets possess at various ranges and indicates the lethal range of those bullets against unprotected targets. An unprotected target is defined as a soldier wearing normal military uniform and equipment without a steel helmet or body armour.

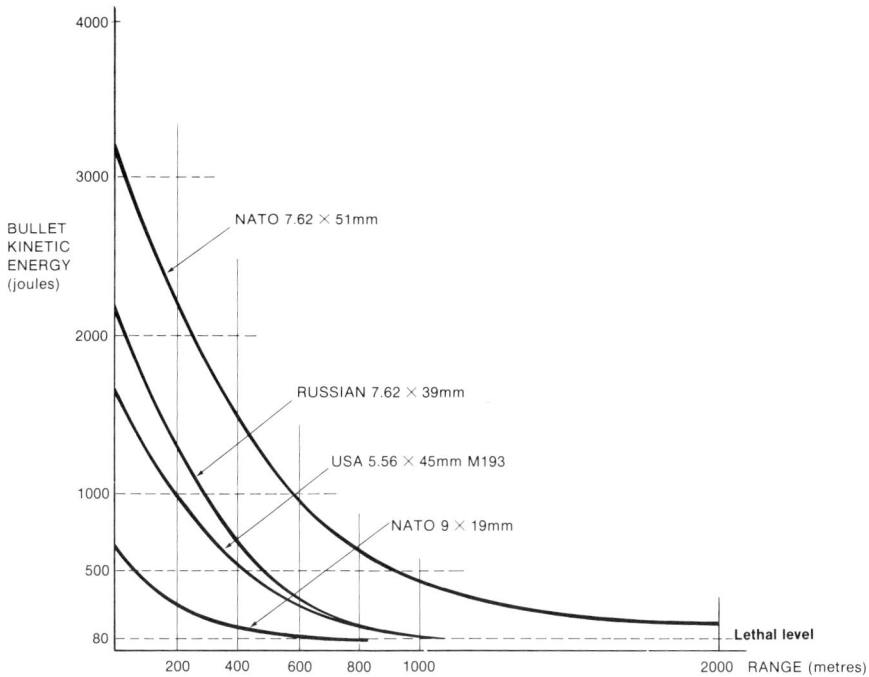

Fig 3.3 Remaining Kinetic Energy versus Range

The following points should be noted from the graph. Firstly the NATO 7. 62 x
51 mm bullets retain energies greater than 80 Joules out to 2000m. Secondly
the Soviet 7. 62 x 39 round which is lighter and has a lower muzzle velocity than
the NATO 7. 62 round has a much reduced lethal range. This round's lethal range
is matched by the 5. 56 x 45 mm M193 round. Possession of more than 80 Joules
is not, by itself, a true minimum energy limit because the bullet must also be
capable of transferring this level of energy to the target. At battle ranges small
arms bullets tend to have considerably greater energies than 80 Joules so that
even if they pass straight through the body there is likely to be a transfer of energy
greatly in excess of 80 Joules.

Defeating the Target

Incapacitating the unprotected man is relatively simple. However once the target
takes cover, or wears protective clothing, incapacitation is more difficult. The
following table indicates the strike energy needed by bullets of two different
calibres to defeat a variety of targets. The figures assume that the bullet strike
is perpendicular to the face of the material.

TABLE 3.1 Approximate Energy Required to Penetrate
Various Objects

Target	Approximate Energy needed by	
	5.56 x 45 mm M193 ball round (Joules)	7.62 x 51 mm NATO ball round (Joules)
Unprotected man	80	80
9" timber	150	200
Soft skinned vehicle	150	200
Steel helmet, or very light armour	420	770
15 mm dural (aircraft)	1500	1800
2" (50 mm) concrete slabs	1200	1500
$4\frac{1}{2}$" (120 mm) brick	2500	3000

Note (1). These figures do not include the extra 80 Joules needed to incapacitate a man protected by these materials.

It can be deduced from this table that as calibre increases the energy required to take the bullet through cover increases. In other words, the larger the bullet the larger the hole it has to punch in the material. It is also clear that the M193 round which has only 1800 Joules at the muzzle can never be expected to pass through a $4\frac{1}{2}$ inch house brick. The NATO 7.62 mm round has the required energy of 3000 Joules out to about 100m. Theoretically it would be possible to produce 5.56 mm bullets capable of penetrating brickwork but the increase in mass could lead to stability problems and increasing their velocity much above 1000 m/sec could cause excessive wear. Recoil and weapon weight could also increase to undesirable levels.

Most soldiers are likely to be wearing steel helmets on the battlefield so that the incapacitation requirement is often coupled with a demand for some armour piercing ability. Approximately 3 mm of mild steel is used in tests to represent the material of the present day steel helmet. A bullet capable of penetrating 3 mm of mild steel will also be able to pierce full body armour comfortable enough to be worn in combat. The requirement for penetration of materials such as sand and wood which may be used in field fortifications is usually relinquished in favour of retaining an easily handled weapon. The energy level needed to penetrate the steel helmet is used to determine the maximum lethal range of the ammunition. For high velocity streamlined projectiles this range tallies well with the range limits of tracer burn out and hit probabilities.

Since the lethality requirement is fundamental to the design of a weapon and its ammunition, the soldier must clearly define the target and its range. Sometimes a combination of targets and ranges is specified. For example, a weapon may be required for use against an unprotected man at 1000m and a protected man at 600m. These criteria should be based on tactical factors such as weapon role, normal fields of fire in a particular terrain and the probability of using supporting fire from other weapons such as mortars or artillery.

WEAPON RANGES

Rifles

Analysis of rifle engagements in World War II, Korea and Vietnam has established that rifle fire is restricted, in action, to ranges much less than usually believed. At Fig. 3.4 is a cumulative frequency graph from which it may be seen that:

30% of all engagements occured at ranges of 100m or less.

70% of all engagements occured at ranges of 200m or less.

90% of all engagements occured at ranges of 300m or less.

95% of all engagements occured at ranges of 400m or less.

It should be noted that these figures have been arrived at by averaging out the short range night engagements with longer range engagements by day. Also that these figures do not show the variations in engagement ranges due to differences in terrain.

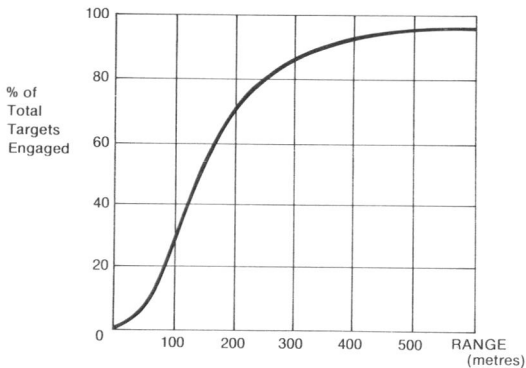

Fig 3.4 Ranges of Rifle Engagements

One can question the need for a rifle which is lethal against the few targets that are over 400m away. It was demonstrated in the NATO Small Arms Trial that it is possible to produce effective fire out to 400m using light, small calibre bullets. These can be fired from light weapons which have little recoil. As the bullet can be made to follow a flat trajectory the overall hit chance of such a system is high.

Machine Guns

The range requirements for machine guns are less easily defined. The medium machine gun (MMG) is often required to provide covering fire between defensive locations which may be up to 2000m apart. The target is likely to be a group of men out in the open and though they may be wearing helmets the larger part of their bodies is unprotected so that they are a relatively easy target to defeat. Therefore a bullet which has little over 80 Joules residual energy is likely to be effective. To reach 2000m the bullet needs to be heavy and, as shown later in this chapter, this is likely to lead to calibres greater than 7 mm.

The section machine gun is often referred to as a light machine gun (LMG). The LMG is required to produce a high volume of fire at relatively short ranges. The target can either be assaulting infantrymen or a dug in enemy. Most of an assaulting infantryman is unprotected and so an 80 Joules residual energy level should prove adequate. The dug in enemy, on the other hand, presents a smaller target much of which is protected by a steel helmet. The range requirement for LMG fire at protected and unprotected infantry varies between armies. Where the range requirement is similar for both LMG and rifle fire, at protected targets, then both weapons can use the same calibre ammunition. However if the LMG range is extended then the logic in favour of separate rounds of different calibres is reinforced.

Some armies use the same weapon for both LMG and MMG roles. The weapon is often termed the General Purpose Machine Gun (GPMG). The photograph at Fig. 3.5 of the US M60 is an example of a GPMG. It is shown in its LMG role; a tripod mounting is required to convert it to an MMG.

Fig 3.5 US M60 GPMG

The calibre of GPMGs tends to be decided by the MMG long range requirements. MMGs make extensive use of tracer rounds to help identify the fall of shot. The trajectory of large calibre rounds is relatively unaffected by the slow combustion of the tracer composition so GPMGs tend to have calibres of around 7-8 mm.

Carrying Power and Stability

The scientific factors which govern bullet range must now be considered. Muzzle velocity is obviously important as is the resistance to motion or drag forces which slow the bullet. The amount of drag is strongly influenced by the bullet's shape and size.

The carrying power of a bullet can be mathematically expressed by the following formula.

$$Co = \frac{1}{Ko} \frac{m}{d^2}$$

Where:

Co = Standard Ballistic Coefficient (Carrying Power)

m = Bullet mass

d = Bullet diameter

Ko = Shape and steadiness factor

The higher the value of Co the farther the bullet will travel as long as the muzzle velocity remains the same. Therefore the designer tries to achieve a high value for m/d^2. For accurate shooting the bullet needs to be stable in flight. In the case of the spin stabilised bullet it is possible to establish the following relationship between stability, bullet shape and rate of spin:

$$S = \frac{A^2 N^2}{4}$$

Where:

S = Stability Factor

A = axial moment of inertia

B = transverse moment of inertia

N = spin rate

μ = a shape sensitive aerodynamic coefficient

Bullets should have a stability factor value of between 1.6 and 2. To obtain a suitable S value the three factors A, B and μ must be carefully balanced. The value of μ increases rapidly with any increase in bullet length and to a lesser extent with increases in bullet diameter. High length to diameter ratios (l/d) will produce low A values and high B values. This will result in low S values which can be compensated for by increasing the spin rate N. However, there is an upper limit to the spin rate because if it is too high there is a possibility that both bullet and bore may be damaged. This, in turn, places limits on the practicable l/d ratio.

For stability the ideal bullet should be short and fat but this would give a low m/d^2 which would result in low carrying power. A compromise is reached with an l/d ratio of around 3-5.

From internal ballistic measurements it has been found that heavy bullets absorb more energy from the propellant. External ballistic work shows that as heavy bullets travel through the air they retain this energy better than light bullets. This is shown in Table 3.2 where the calibre and muzzle velocity have been kept constant and the weight of the bullet increased.

TABLE 3.2 Variation of Performance with Changing Bullet Weight

Bullet Weight (grams)	7.1	8.4	9.7
Range at which a specific target can be defeated (metres)	590	660	730

Most bullets are made of a combination of materials. The density of a bullet depends on the proportions of these materials. The densities of some of the materials which may be included in a bullet are shown below.

TABLE 3.3 Densities of Bullet Materials

Material	Density gms/cc
Depleted Uranium	18.7
Lead	11.3
Copper	8.9
Gilding metal (90% Cu:10% Zn)	8.8
Steel	7.8
Aluminium	2.7
Phosphorous	1.8
HE/Tracer Composition	1.6

Though tracer rounds are longer than ball rounds they are still lighter. It can be seen in Fig. 3.6 that the densities and lengths of the 7.62 x 51 mm ball and tracer rounds differ greatly. Only the bullet diameter and nose shape are the same.

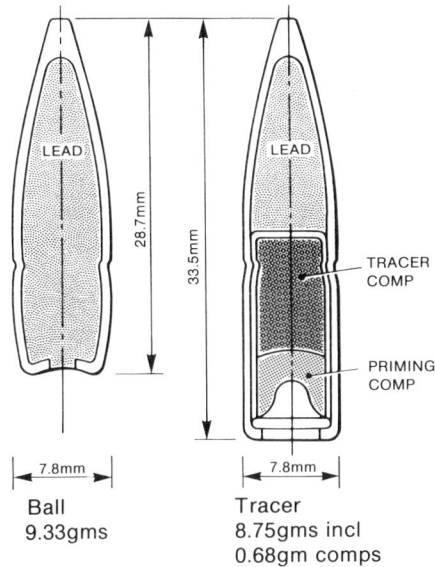

Fig 3.6 NATO 7.62 × 51mm Bullets

As the tracer composition is consumed the bullet weight changes. Thus the carrying power and stability factor of a tracer are continuously changing in flight. It is not surprising therefore, that it is impossible to match the trajectories of ball and tracer rounds. The best that can be hoped for is a minimal mismatch at the normal engagement ranges where tracer is needed to observe the fall of shot.

The upper limit for bullet weight is governed by the following facts. Firstly, there is the need for a tracer round to be longer than the ball round: the length of any round being governed by its stability ($1/_d$ less than 5). Secondly, the reduced stability of heavier, and hence longer bullets, necessitates an increased twist of rifling, which itself can lead to problems of wear.

The requirement for bullets with the power to penetrate steel helmets at long range has resulted in small calibre bullets themselves being made with steel tips or cores. This trend is demonstrated by the following table which shows the composition of various small arms bullets.

TABLE 3.4 Bullet Construction

Bullet	Composition
.22 LR	Lead, bare
7.62 x 51 mm NATO	Lead core, jacketted (1)
5.56 x 45 mm M193	Lead core, jacketted
5.56 x 45 mm SS109	Steel tip + lead core, jacketted
5.45 x 39 mm AK74	Lead tip + steel core, jacketted

Note (1)   Bullet jackets are normally made from gilded metal or copper coated steel.

HIT PROBABILITY

Certain aspects of hit probability are relevant to the choice of weapon calibre. The three most important are vertex height, wind deflection and recoil. These will be covered in turn starting with vertex height. Other aspects of hit probability will be covered in Chapter 5.

Vertex Height

If air resistance is ignored and gravity is considered to be the only force acting on a bullet during its flight, then the distance it falls can be expressed as:

$$S = \tfrac{1}{2} gt^2 \qquad \text{Where } S = \text{drop}$$
$$g = \text{acceleration due to gravity}$$
$$t = \text{time of flight}$$

Sight setting takes this into account so that the bullet is fired at an upward angle instead of parallel to the earth. Vertex height is the distance between the ground and the apogee of the trajectory. The diagram at Fig. 3.7 shows the relationship between the line of sight from firer to target and the path taken by the bullet after adjustment for drop. The diagram has been deliberately distorted so as to magnify the difference between the two.

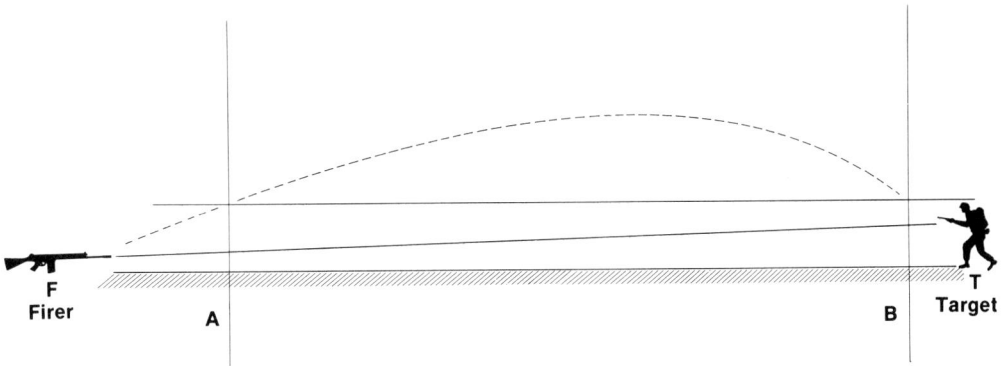

Fig 3.7 Relationship between the Trajectory Height and the Line of Sight

The solid line from T to F is the line of sight of the weapon. The dotted path is the trajectory of the bullet. If the target were to move towards the weapon, then although it would always be in the sights of the firer, it would not be hit between points A and B because the bullets would pass overhead. Vertex height can be reduced by increasing muzzle velocity. Low vertex height is important for modern weapons which have a single battlesight setting for ranges under 300-400m. If it is the firer's intention to deny ground to an enemy then it is useful to keep the bullet's trajectory well below a man's height. Air resistance not only slows a bullet's horizontal movement but also slows its vertical drop. When air resistance is taken into account the amount that a bullet drops at short ranges is shown in Fig. 3.8.

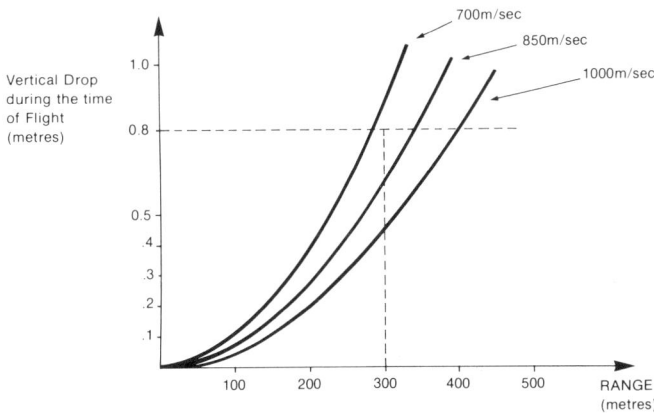

Fig 3.8 Vertical Drop at Various Ranges for Bullets of Different Muzzle Velocities

If an average human target is assumed to be 1.6m high and the point of aim is central, then there is a 0.8m margin for vertical error. Figure 3.7 shows, therefore, that for weapons firing bullets with muzzle velocities in excess of 750 m/sec a single sight setting, for ranges up to 300m, should be adequate. It also shows why correct range estimation becomes more important as muzzle velocity decreases.

Wind Deflection

Cross winds deflect a bullet from its true trajectory. The amount of deflection depends not only on the wind direction and speed but also on the size of the bullet and its time of flight.

Wind deflection is at its maximum when the wind is at right angles to the bullet trajectory. Oblique winds, therefore, cause less deflection. The more time the wind has to act on the bullet the greater its effect will be. Thus the amount of deflection increases with range. For similar reasons high velocity bullets are less likely to be deflected than slow ones. The smaller the surface area of a bullet the less it should be deflected but this fact tends to be countered by small bullets having a reduced mass. Light bullets are more easily deflected than heavy ones. As a broad generalisation high velocity, heavy bullets are least likely to be affected by cross winds.

Recoil

It is accepted that the amount of recoil affects the chance of hit when firing bursts. There is less conclusive evidence to suggest that the same is true of single shot fire. The bullet is still just within the bore when the weapon starts to recoil. Correct zeroing should be able to eliminate this effect but the zero has to be changed for each fire position. There is no time for this in battle. This aspect is covered more fully in Chapter 5. The soldier is also less likely to flinch when firing if he is not expecting an uncomfortable blow in the shoulder. Therefore low recoil will improve hit probability.

The relationship between weapon weight and recoil can be established by considering the change in momentum of the weapon and bullet up to moment it leaves the barrel.

In Chapter 2 it was shown that recoil can be expressed mathematically as:

$$R = \frac{1}{2} \frac{(mv)^2}{M}$$

Where R = recoil energy

M = weapon mass

m = bullet mass

v = muzzle velocity

This was a simplification and recoil can be more accurately expressed when the mass of the propellant gases is taken into account. These increase the forward momentum of the system. The formula can then be rewritten as:

$$R = \tfrac{1}{2} v^2 \frac{(m + kc)^2}{M}$$

Where c = charge mass

k = a constant relating to the efficiency of the system. It normally lies between 1.5 and 1.7 and is found by experiment.

The combination of bullet mass and velocity values tends to be high. These values are decided by the tactical requirement to penetrate a specific target at a specific range. The soldier wants a light weapon with low recoil. It can be seen, then, that the above formula is not easily balanced. Table 3.5 shows the recoil values of a number of weapons.

TABLE 3.5 Free Recoil Energy of Various Weapons

Weapon	Round Size mm	Weapon Mass kg	Bullet Mass gm	Charge Mass gm	Muzzle Velocity m/sec	Weapon Recoil Velocity m/sec	Free Recoil Energy J
Double Barrel Shotgun	12 Bore G Prix	3.00	33.14 incl wads	2.64	330	4.08	25.0
No 5 Rifle	7.7x56R	3.20	11.49	2.40	729	3.39	18.4
No 4 Rifle	7.7x56R (.303)	4.07	11.49	2.40	748	2.74	15.3
L1A1 Rifle	7.62x51	4.57	9.33	2.85	838	2.48	14.0
AKM	7.62x39	3.45	7.97	1.62	715	2.15	8.0
AR 15	5.56x45	3.04	3.56	1.56	1000	1.94	5.7
Ingram SMG	11.4x23 (.45ACP)	3.19	15.16	0.32	280	1.37	3.0
AK 74	5.45x39	3.45	3.45	1.37	900	1.44	2.5
L2A2 SMG	9x19	3.21	7.45	0.39	390	0.97	1.5
HK 33 with insert	.22LR	4.3	2.6	0.10	310	0.23	0.1

Conflicting Influences on Hit Chance

It is not easy for the soldier to make a clear technical statement on chance of hit.
The following facts conflict; firstly, high velocity bullets increase hit chance
because their flat trajectory makes exact range estimation and sight setting less
critical; secondly, light weight and low velocity bullets increase hit chance
because the light weight, low recoil rifles which use them enable the firer to
maintain a steady aim; thirdly, heavy and high velocity bullets are less easily deflected by wind and brush and therefore have a better chance of hit. It is simpler
for the soldier to base his hit chance requirement on the performance of current
equipments rather than make a statement in numerical terms. For example, the
hit chance requirement for a rifle and its ammunition could be expressed in the
following way; the vertex height should be no greater than the NATO 7.62 x 51 mm
round; the recoil should be similar to that of the US Armalite rifle; the wind deflection should be less than that of the 5.56 x 45 mm M193 round.

EXPANSION RATIO

There is one more important criterion which influences the choice of calibre.
This is the weapon's expansion ratio. The expansion ratio is defined as:

$$\frac{\text{Volume of the barrel} + \text{volume of the cartridge case}}{\text{Volume of the cartridge case}}$$

It is a measure of the volume available for the burning of the propellant gases.
If the value is too low then muzzle pressure is too great, efficiency is reduced
and there is excessive flash. The UK No 5 rifle used in World War II had a barrel 163 mm shorter than the No 4 but fired the same round. The photograph at
Fig. 3.9 shows the large flash eliminator which had to be fitted to the No 5 rifle.

Fig 3.9 The UK No 4 and No 5 Rifles

A requirement limiting barrel length in favour of easy weapon handling in confined spaces such as armoured vehicles and buildings puts an upper limit on the expansion ratio (ER). Most modern weapons have an ER value of 8 though it is practicable to reduce this to 7 provided that the weapon is not expected to produce a high volume of fire.

For a given barrel length and expansion ratio the case volume and calibre are narrowly defined. Assuming that the expansion ratio remains unaltered, as the calibre decreases, the case volume must decrease. This leads to a reduction in charge weight which in turn leads to lower muzzle energy. The expansion ratio therefore fixes the upper limit of muzzle energy for a particular calibre.

The sketch at Fig. 3.10 shows the relationship between recoil energy, flatness of trajectory, wind deflection and expansion ratio.

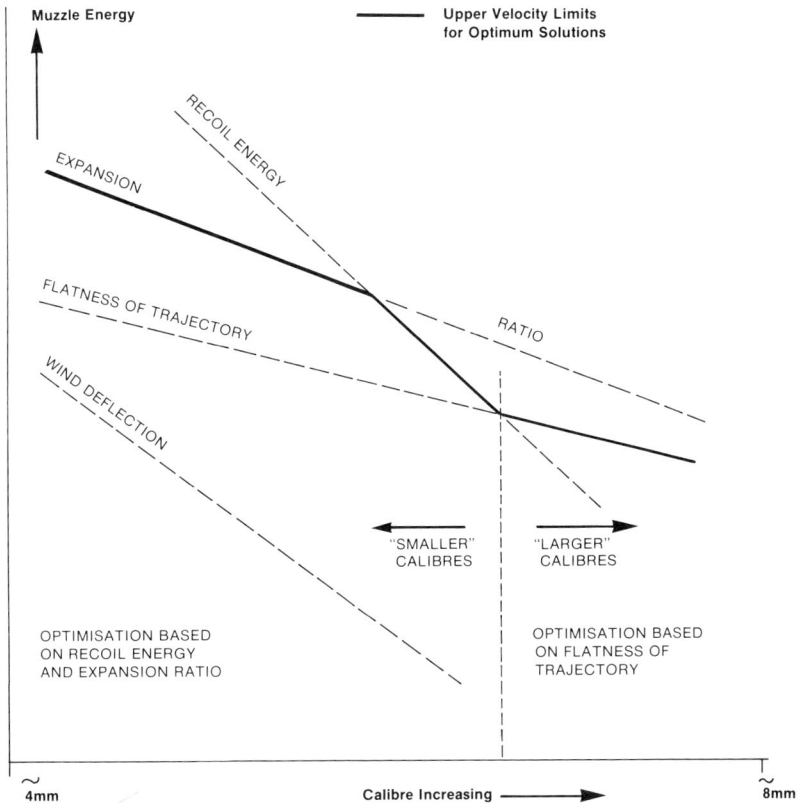

Fig 3.10 Factors Affecting Muzzle Energy

The position of these four factors can vary but this graph shows the relationship between them when based on the details shown under the sub-heading Conflicting Influences on Hit Chance. When considering smaller calibres it can be seen that wind deflection and flatness of trajectory indicate the lowest acceptable level of muzzle energy. Also, in this case, wind deflection is a less severe constraint than flatness of trajectory. The upper limits of muzzle energy are set by the acceptable recoil and ER (barrel length). If barrel length and recoil are allowed to increase then greater muzzle energies are practicable and the bullet will then be effective at longer ranges. It can be seen that at larger calibres it is impossible to achieve the muzzle energies needed for the required flat trajectory from a low recoil weapon. The relationship between these criteria would have to be altered by adjusting the values accorded to each.

SUMMARY

In weapon design it is important to establish the effect required at the target.
This leads to a need to define the target and the degree of protection it is likely to have on the battlefield. From such an analysis it is possible to decide the level of strike energy needed by the bullet. This energy level varies not only with calibre but also with the type of protection. The next important factor is the range at which the bullet must be effective. From this it is possible to work out the necessary muzzle velocity for a particular design of bullet. The muzzle velocity of the bullet will be governed by such additional military factors as flatness of trajectory, recoil, acceptable wind deflection and weapon length. A compromise may have to be reached as the soldier's ideal requirements are often difficult to achieve, particularly for the larger calibres.

SELF TEST QUESTIONS

QUESTION 1   What factors influence the lethality of a bullet?

Answer   ....

....

QUESTION 2   What factors influence the penetration of a bullet?

Answer   ....

....

QUESTION 3   What is the maximum effective range of a 7.62 x 51 mm bullet?

Answer   ....

....

QUESTION 4   What are the advantages of using a heavy bullet?

Answer   ....

....

QUESTION 5   What are the disadvantages of using a large calibre bullet?

Answer   ....

....

QUESTION 6   What are the problems caused by having a long bullet?

Answer   ....

....

QUESTION 7   What are the advantages and disadvantages of short barrels?

Answer   ....

....

QUESTION 8   What are the problems with tracer rounds?

Answer   ....

....

QUESTION 9   In what order should the main factors affecting bullet performance
be considered?

Answer   . . . .

. . . .

QUESTION 10   What factors conflict in the choice of calibre for a GPMG?

Answer   . . . .

. . . .

ANSWERS ON PAGE 181

4.

Heating

INTRODUCTION

When a weapon is fired heat is produced from two sources; a little comes from the friction between the bullet and the bore and the rest from the burning propellant gases. Some of this is absorbed initially by the barrel and then spreads slowly throughout the weapon but most of the heat flows to the outer surface of the barrel. The barrel can lose heat to its surroundings both by convection and radiation. However barrel heat tends to build up because heat absorption is more rapid than heat loss.

Theoretically the barrel could become so hot that it would be impossible for any more heat to be absorbed. This is because the heat transferred from bore to gas would equal the heat transferred from gas to bore. However, almost certainly, before this stage is reached the weapon would become too hot to hold, be damaged and fire inaccurately.

This chapter examines the effects of overheating on small arms and the ways in which the problems can be minimised. The process of heat flow is covered first.

HEAT INPUT

The two sources of heat are the hot propellant gases and the friction between the bullet and the bore. It is difficult to establish accurately the amount from each source. Trials have been carried out with rounds likely to produce different amounts of friction. After each round had been fired a similar number of times barrel temperatures had risen by approximately the same amount. This supports the view that the bulk of the heat comes from the propellant gases.

The propellant flame temperature is approximately 2000°C. Heat is transferred to the bore by convection and radiation over a 5 millisecond period after cap strike. During this time the bore surface temperature can reach to between 700°C and 1000°C. Once the hot gases have dispersed the bore temperature drops but at a much slower rate.

Figure 4.1 shows the temperatures of the inner and outer surfaces of a barrel after a number of rounds have been fired.

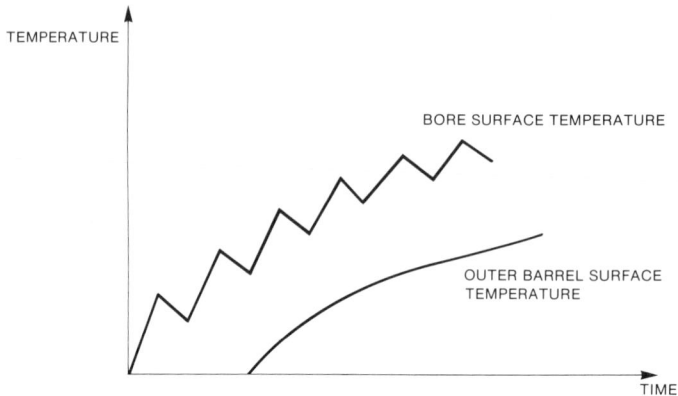

Fig 4.1 Bore and Barrel Surface Temperatures

The amount of heat produced by different rounds depends mainly on the amount of propellant (charge mass), the rate at which the propellant burns and the flame temperature. The amount of this heat subsequently transferred to the barrel depends on the type of barrel material, the composition of the propellant gases, the area of the bore and the temperature difference between the bore and the hot gases. As the bore surface heats up, this temperature difference decreases, heat input is reduced and the temperature rise becomes progressively smaller.

As will be shown later, the cooling rate of the bore surface is very slow in relation to the rate of heating. It is so slow that its effect at normal battlefield rates of fire is negligible and can be ignored. Subsequently all graphs in this chapter will show a continuous heat line rather than the, more strictly accurate, jagged one.

Heat input is greatest just after the commencement of rifling and gradually decreases towards the muzzle. The greater the thickness of the barrel walls, the longer it takes for heat to reach the outer surface. Barrel thickness normally decreases towards the muzzle. The outer surface temperature is therefore unlikely to be uniform along the barrel and is only a guide to the bore temperature which is some $30^{\circ}C$ higher.

If the bore is kept at a high temperature over a long period of time the weapon will become damaged. Therefore it is important to study ways in which heat can be lost. The next section of this chapter indicates how a barrel loses heat.

HEAT FLOW FROM THE BORE

Metals are good conductors of heat whilst liquids and gases are poor conductors. Heat is conducted relatively quickly to the outer barrel surface and gradually to all other parts of the weapon, unless these are insulated. Once heat reaches the outer surface of the weapon it can only be lost through convection and radiation which are slower processes than conduction.

Conduction

It can be shown experimentally that the rate of conduction of heat through a cylinder is directly proportional to the surface area and to the temperature difference between the inner and outer surfaces; it is also inversely proportional to the cylinder wall thickness. Theoretically, therefore, the rapid conduction of heat from the bore to the outer surface of the barrel is best achieved by making the barrel walls very thin. Even if this were practicable, the heat loss to the surroundings from the barrel surface would remain so slow that the weapon would still overheat. To avoid this the barrel must become a heat reservoir and this can only be achieved by making the walls thick. In practice rifle barrels which are thick enough to withstand rough handling and firing pressures are also thick enough to provide an adequate heat reservoir at reasonable battlefield rates of fire. Weapons such as LMGs and MMGs, which are used to give high volumes of automatic and sustained fire, need to have thickened barrels to cope with the greater heat produced.

Convection

When a gas or liquid touches a hot object, as in the case of air in contact with a hot barrel, some heat is conducted to the layer next to the object. The density of this layer then decreases and so it rises and its place next to the hot object is taken by a cooler layer of fluid. This process is called natural convection. If any outside assistance, such as a fan, is used to increase the speed of movement of the fluid layers the process is called forced convection and is a more efficient method of exchanging heat.

The rate of loss of heat by convection can be found experimentally to depend directly on the temperature difference, the area of the hot object and on certain properties of the fluid. Liquids, for example, are much better at transferring heat than gases.

Radiation

The third method of losing heat is by radiation. Thermal radiation occurs in the form of electromagnetic waves. All objects give and receive radiated energy. When an object is hotter than its surroundings it is a net emitter of radiation. Experimentally, heat loss by radiation has been found to depend on the temperature of the hot object and on certain of its physical properties, namely its area and its emissivity. Materials which absorb heat easily also emit heat easily.

Materials which reflect heat neither absorb nor emit heat well. A barrel made of oxidised metal has higher emissivity than one with a shiny surface; it therefore radiates heat more efficiently. Heat loss by radiation becomes progressively more efficient as the temperature of the hot object rises because the loss rate depends on the fourth power of the absolute temperature (T^4).

Heat Transfer in Small Arms Barrels

It has been demonstrated experimentally that the average heat intake for high velocity weapons increases as calibre increases. This is logical since the surface area of the bore is greater and the amount of propellant normally increases with calibre. The following table gives the average heat input of two different calibre rounds after 100 and 200 rounds have been fired. It can be seen that heat input decreases as the barrels get hotter.

TABLE 4.1 Heat Input from Two Rounds

Round (mm)	Charge Weight (grams)	Average Heat Input per Round (Joules)	
		First 100rds	Second 100 rds
7.62x51	2.9	2200	2000
5.56x45 M193	1.6	1400	1100

The following graph gives an indication of the temperatures reached by barrels of weapons firing the two rounds mentioned in Table 4.1.

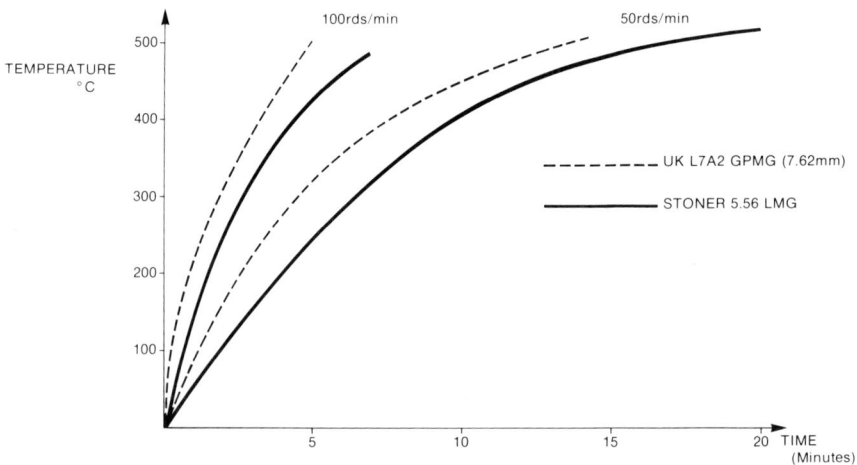

Fig 4.2 Temperatures of Weapon Barrels

The potential heat loss from a weapon can be calculated mathematically. In the examples quoted below the weapon has a barrel 51 cms long with an external diameter of 20 mm. These measurements are representative of both 5.56 mm and 7.62 mm section weapons. When the barrel temperature is $200^{\circ}C$ and the ambient temperature $15^{\circ}C$ then the theoretical heat loss from convection is 39 watts and from radiation is 57 watts making a total of 96 watts. To maintain a heat input of 96 watts the rate of fire from a 7.62 x 51 mm weapon needs to be 3 rounds per minute. A temperature of $200^{\circ}C$ can be achieved at very modest rates of fire, particularly in the case of machine guns. On the battlefield, $200^{\circ}C$ is not only easily achieved but is bound to be surpassed as the weapon is almost certain to be fired at greater than 3 rounds per minute.

When the barrel temperature reaches $500^{\circ}C$ the theoretical heat loss through convection is 150 watts whilst that through radiation has increased to 506 watts. This tenfold increase over the $200^{\circ}C$ rate shows that radiation becomes the dominant heat loss process as temperature rises. The total heat loss is 656 watts which is the amount of heat produced by firing a 7.62 x 51 mm weapon at 18 rounds per minute. On the battlefield $500^{\circ}C$ is normally only achieved by machine guns firing at intense rates which are far in excess of 18 rounds per minute. However, once reached, a barrel temperature of $500^{\circ}C$ is easy to maintain as 18 rounds per minute is a modest rate of fire for a machine gun. These examples emphasise the fact that heat input to small arms barrels is far in excess of heat loss by natural convection and radiation. Meeting the battlefield volume of fire requirements makes it difficult to design weapons which do not suffer from overheating problems.

OVERHEATING PROBLEMS

The high temperatures reached by small arms barrels have a number of effects; firstly parts of the weapon become too hot to handle, secondly the weapon may be damaged and fire inaccurately and thirdly the ammunition may ignite prematurely. These factors will be dealt with in turn starting with premature ignition which is usually called cook off.

Cook Off

Propellant is normally ignited by the flames which are produced by the cap in the base of the cartridge after it has been struck by the firing pin. Cook off occurs when the propellant reaches its ignition temperature by absorbing heat from the chamber walls. Ignition temperatures for small arms propellants lie in the range $180^{\circ}C$ to $200^{\circ}C$. However, cook off does not happen immediately after the chamber reaches ignition temperature. The brass cartridge case gives the propellant some protection because it takes time for the heat to flow from the chamber walls through the brass case. Modern self loading weapons eject the brass case within milliseconds of firing so that there is little time for heat to pass into the chamber from the case. The chamber gains its heat indirectly by gradual conduction from the barrel. Therefore it takes some time for the chamber to reach ignition temperature and then for that heat to reach the propellant. The following table gives an indication of the time delays involved.

TABLE 4.2 Cook Off Times for a 5.56 mm LMG

Rate of Fire Rds/Min	No of Rounds Fired	Time (Mins)	Temperature of Barrel °C	Cook Off (Secs)
240	240	1	350	None
120	360	3	400	16
120	300	$2\frac{1}{2}$	380	38
80	320	4	380	42
80	280	$3\frac{1}{2}$	375	35
80	240	3	340	None

Cook off is not a problem, even for weapons likely to become very hot, as long as the round is only momentarily chambered. It is for this reason that machine guns fire from open bolts, which means that the working parts are held to the rear until the trigger is pressed. Operating the trigger causes the working parts to move forward, collect a round from the feed system, chamber it and automatically fire it. Only when there is a malfunction does the unfired round remain in the chamber for more than a few milliseconds. Good weapon handling drills will prevent it staying there long enough to cook off.

Open bolt weapons are less accurate than those firing from a closed bolt. The reasons for this are explained in Chapter 5 which covers Hit Probability. Rifles are required to be very accurate and so they must fire from a closed bolt. This makes it possible for an unfired rifle round to remain in the chamber long enough for cook off to occur. In the case of closed bolt weapons an upper temperature limit must be established. The factors which should be taken into account are firstly, battlefield rates of fire, secondly, the time taken for the chamber to reach propellant ignition temperature and thirdly, can the risk of cook off under some circumstances be militarily acceptable. A temperature of 250°C on the outer barrel surface just forward of the commencement of rifling is one such criterion. If a weapon reaches a temperature of 250°C and maintains it for only a few seconds before cooling to below 200°C then cook off is unlikely to occur. If, however, the barrel is maintained at greater than 250°C for over a minute there is a strong probability that cook off will occur if a round is held in the chamber.

Wear and Erosion

Confusion can arise over the use of the terms wear and erosion. Wear is the result of mechanical abrasion which removes layers of metal from the bore. Erosion is the removal of particles of the bore by the hot propellant gases. The gradual wear which takes place in all barrels was described in Chapter 2; erosion and rapid wear take place at high barrel temperatures. Severe barrel wear and erosion causes the bullet to deviate from the line of sight of the weapon and in extreme cases to become so unstable that it travels broadside on. Figure 4.3 shows the holes made in a target at 25m by bullets fired from a shot out 5.56 mm LMG barrel.

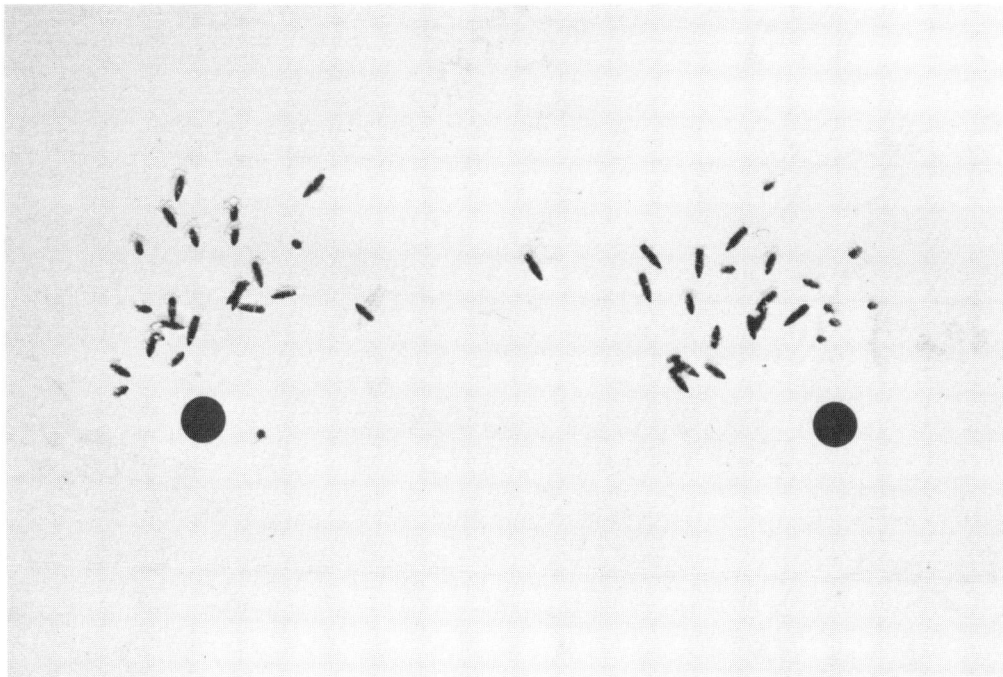

Fig 4.3 Bullet Holes from a Shot Out Barrel

Inaccurate Fire

A purely transient effect of high barrel temperature is the expansion of the bore diameter by .028 mm per $100^{o}C$. This can make it difficult for the bullet to set up. Set up is a process which is covered in more detail in Chapter 5. If the fit between bullet and bore is loose at the muzzle then the bullet is likely to emerge with unacceptable yaw. The overall effect is to increase the dispersion of the shots around the target.

A further transient effect of high barrel temperature is a weakening of the strength of the barrel steel which may lead to permanent deformation and damage. The graph at Fig. 4.4 shows how the hardness of a typical barrel steel decreases with increased temperature. The dotted horizontal line shows the hardness of a typical bullet jacket at ambient temperature.

It can be seen that with a barrel at approximately $600^{o}C$ the cold bullet jacket is harder than the lands of the rifling. The large mechanical forces which occur at the commencement of rifling will thus cause the softer lands to be worn away. The onset of this damage is sudden and permanent.

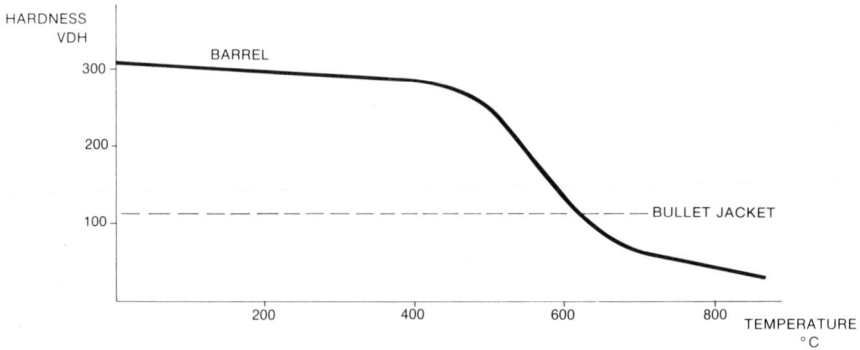

Fig 4.4 Variation of hardness with temperature

When the lands have been worn away at the commencement of rifling the bullet is given less spin than is needed for the desired ballistic stability. Additionally, the bullet must take longer to reach the rifling and will therefore meet it at a higher speed. This places greater stress on both bullet and lands.

As the hardness of the barrel steel decreases, so the effects of the pressure of the propellant gases become more noticeable and in extreme cases the barrel may bulge near the commencement of rifling where the pressure is greatest. The deformation of this part of the barrel makes bullet set up more difficult and causes muzzle velocity to drop. A further effect is the escape of gas past the bullet before set up occurs and in extreme cases for much of the time that the bullet is in the bore. The escape of hot high pressure gas through a narrow aperture causes intense localised heating which may melt the bore surface. This process is termed gas wash erosion and the damage it does is permanent.

High barrel temperatures accelerate wear and cause erosion. As a result of wear and erosion, bullets emerge at variable muzzle velocities and with reduced spin. Fire from damaged barrels is inaccurate and in extreme cases the bullets become totally unstable. To avoid this, bore temperatures must be kept well below 550°C. Barrel outer surface temperatures should not exceed 400-450°C so as to allow a safety margin. Wear and erosion create permanent damage but the effects of this damage may be less apparent at low temperatures when effective bullet set up is more possible. However as the temperature increases, expansion of the bore through thermal forces and gas pressure reduces the possibility of efficient bullet set up. Fire from the weapon then becomes inaccurate.

Handling

Heat is conducted from a hot barrel to the rest of the weapon. During normal battlefield engagements the process is slow enough to avoid overheating those parts of a weapon which a soldier needs to handle. However some weapons do

have restricted volumes of fire because of handling difficulties though these problems probably only occur in emergency situations. It is not easy to be more
precise about this aspect of overheating. Good design and the use of insulating
materials should minimise any difficulties.

REDUCING OVERHEATING PROBLEMS

There are two basic methods of reducing overheating problems. The weapon can
be designed with special features which either specifically improve the heat loss
from the barrel to its surroundings or the effects of overheating can be minimised.
In the former category are such features as liquid cooling jackets and finned barrel surfaces; in the latter are barrel liners, heavy barrels, cooler propellants
and careful positioning of components such as triggers and handguards. The use
of changeable barrels could fall into either category. These features will be discussed starting with those designed to improve heat loss from the barrel surface.

Finned Barrels

In theory, increasing the surface area of the barrel improves the rate of heat
loss to the surroundings. One method is to increase the diameter and thus the
circumference of the barrel but this is likely to make the barrel too heavy. The
other method is to use fins as shown in Fig. 4.5.

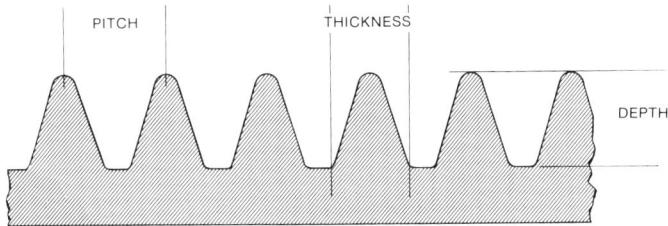

Fig 4.5 Finned Barrels

The fins can be either longitudinal or annular. Their efficiency depends on their
depth, thickness and pitch. These three dimensions are also shown in Fig. 4.5.
The effect each of these three dimensions has on heat dissipation conflicts so that
the designer has to find a compromise. For example, the smaller the pitch the
greater the possible number of fins and consequently the greater the surface area.
However, the smaller the pitch the smaller the air gap so that the fins tend to
radiate heat to each other rather than lose it by convection or radiation to the air.

In practice it is found that unless there is an air flow over the fins there is only
minimal improvement in heat loss. Creating an artificial air flow over a small
arm barrel is difficult. It was tried with marginal success with the World War
I Lewis gun shown in Fig. 4.6.

Fig 4.6 Lewis Gun

When this weapon was fired, cool air was drawn forward through the jacket to equalise the pressure reduction at the muzzle following muzzle blast. Owing to the extra bulk and weight it is most unlikely that finned barrels and forced air convection systems will be found on other than vehicle mounted weapons. Even for vehicle mounted systems, where weight is not such an important criterion, there are other more effective systems for avoiding overheating.

Water Jackets

Another method of improving heat loss is to surround the barrel with a liquid rather than a gas. Not only are liquids better than gases at absorbing heat by conduction and convection but they also absorb even more heat if allowed to vaporise. The liquid normally used is water; not because of any particular heat absorption properties but because of its ready availability.

A well known water cooled weapon is the Vickers Medium Machine Gun shown in Fig. 4.7.

Fig 4.7 Vickers Medium Machine Gun

The water jacket contained four litres which boiled after 600 rounds had been fired at the rate of 200 rounds per minute. If the gun continued to fire the water evaporated at the rate of 0.6 litres every 1000 rounds. Most of the steam was collected in a condensing can and used to refill the water jacket.

The main problems with this system of cooling are the resupply of water, the cracking of the water jacket by ice formation in cold climates, the vulnerability of the jacket to shrapnel and the additional weight of the water jacket and the water. The weight of the Vickers MMG without its 22.7 kg tripod was 18.2 kg of which about a third could be attributed to the cooling system. Liquid cooled guns are unlikely to be found in infantry sections because of their weight but could be used as vehicle mounted or medium machine guns. In view of the problems mentioned earlier it is likely that other methods of cooling will be preferred even for vehicle mounted machine guns.

Heavy Barrels

The great advantage of a heavy barrel is that it acts as a heat reservoir. To
reach a given temperature more heat is absorbed by a heavy barrel than a light
one. Therefore a heavier barrel permits a weapon to be fired at a particular
rate for longer or at a higher rate over a similar period. Heavy barrels should
also increase loss because they have inherently larger surface areas than lighter
barrels made of the same material. However, a barrel with a diameter large
enough to have a significantly increased area would result in an unacceptably
heavy weapon.

The graph at Fig. 4.8 shows the predicted performance of 5.56 mm barrels of
different weights. The predicted performance is based on a computer analysis.
The results at low barrel weights correspond closely to experimental data.

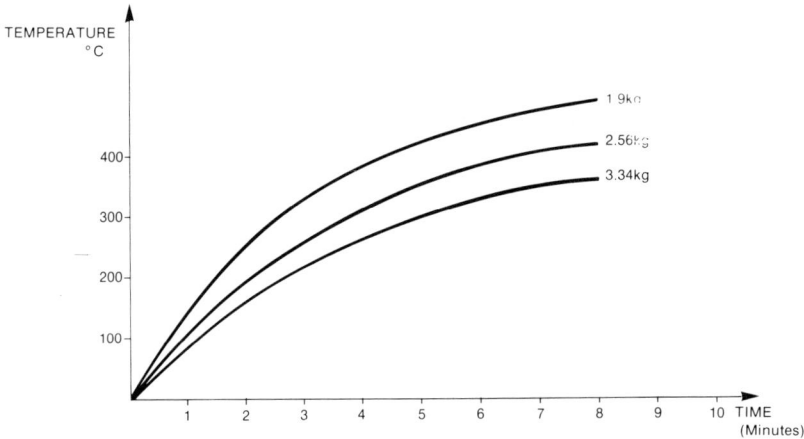

Fig 4.8 Predicted performance of various 5.56mm barrels

Increasing barrel weight seems an attractive method of avoiding overheating but
it does have its drawbacks. Firstly heavy barrels take longer to cool than light
ones. In prolonged firefights the weapon with the heaviest barrel gains least
advantage from any pauses in firing. The graph at Fig. 4.9 shows the rate of
cooling of two barrels of different weights.

A less obvious drawback to heavy barrels is that an increase in barrel weight
leads to a corresponding increase in the weight of other parts of the weapon.
Not only must the mountings be stronger but the breech block has to be heavier
in order to maintain the correct weight ratio with the barrel. Any imbalance in
this ratio can lead to malfunctions in the cycle of operations which are covered
in Chapters 7 and 8.

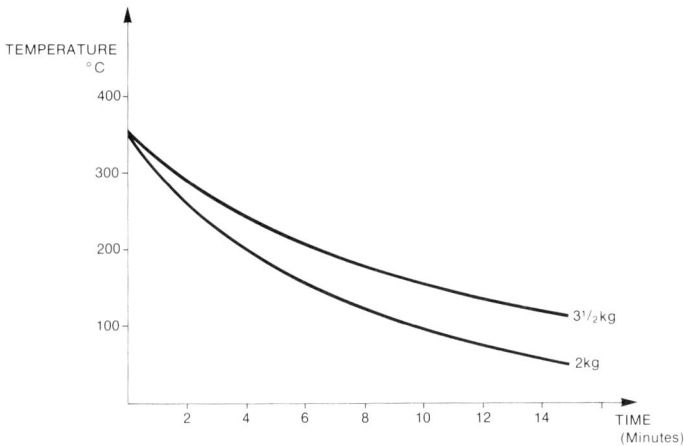

**Fig 4.9 Variation of the Rate of Cooling for
Different Barrel Weights**

In practice the minimum weight of the barrel is governed by the barrel's ability
both to withstand firing and handling stresses and to avoid overheating at expected
battlefield rates of fire. In the case of section weapons the barrel thick enough
to withstand the former is usually heavy enough to cope with the latter. Increasing the weight of the barrel can be an effective means of avoiding overheating
problems providing the gun is either vehicle mounted or only to be carried over
short distances.

Changeable Barrels

Before the barrel reaches a critical temperature it can be exchanged by the firer
for a cooler one. This method is commonly used to avoid the problems of wear
and erosion. However, the ease with which 250°C can be reached and maintained
makes it impracticable to use this method to avoid cook off. The graph at
Fig. 4.10 shows the increase in temperature of a 5.56 mm Stoner LMG firing
from two barrels which have been changed every 250 rounds. Though each barrel
gets progressively hotter a temperature of 400°C is not exceeded even after 20
minutes firing. Figure 4.2 showed that a Stoner LMG which used only one barrel
reached 400°C in 10 minutes when firing at the same rate of 50 rounds per minute.

It is of interest to note the side effect of barrel changing on the temperatures of
the trigger and handguard. When firing at 50rds/min and using only one barrel
the handguard reached 40°C in 10 minutes. If this temperature is exceeded the
handguard becomes uncomfortable to hold.

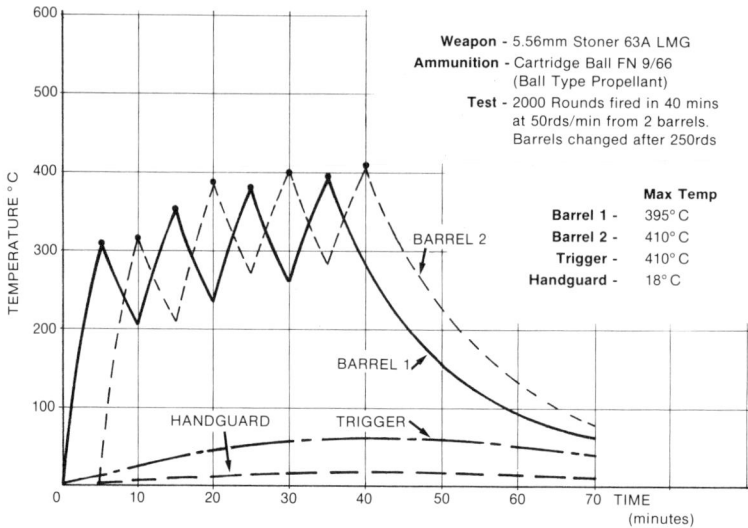

Fig 4.10 Temperature Rise of Stoner LMG using two Barrels

A changeable barrel increases the weight of the weapon since the designer has to include a carrying handle and some mechanism for fixing the barrel to the weapon. Such a weight increase is in the order of half a kilogram.

A fixed single barrel is theoretically better at avoiding overheating problems than two changeable barrels whose total weight is equivalent. However the changeable barrel system does allow the firer the option of carrying only one lightweight barrel if he feels such a risk is warranted, as for example on a night patrol.

Barrel Liners

The effects of wear and erosion can be delayed by increasing the hardness of the bore surface. The graph at Fig. 4.11 shows the increase in high temperature hardness of various materials which are used in the manufacture of barrel liners. The use of the harder materials will delay the onset of wear and erosion.

All modern barrels are given some form of treatment to improve their high temperature strengths. Chromium plating is an example of this. It involves depositing a very thin layer of chromium over the bore surface. This is a costly and difficult process but can improve barrel life significantly.

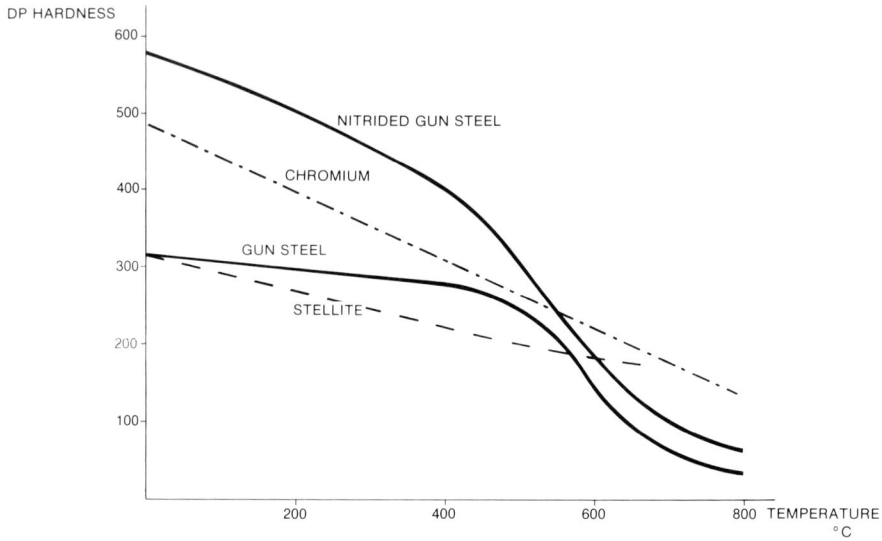

Fig 4.11 Variation with Temperature of the Hardness of various Barrel materials

Another approach is to fit a stellite liner into the first few centimeters of the barrel. Stellite is a very hard alloy composed mainly of cobalt, chromium and molybdenum. The photograph at Fig. 4.12 shows a typical stellite liner. These are used in the US M 60 GPMG and were fitted in the heavy barrels of the UK version of the MAG 58 GPMG.

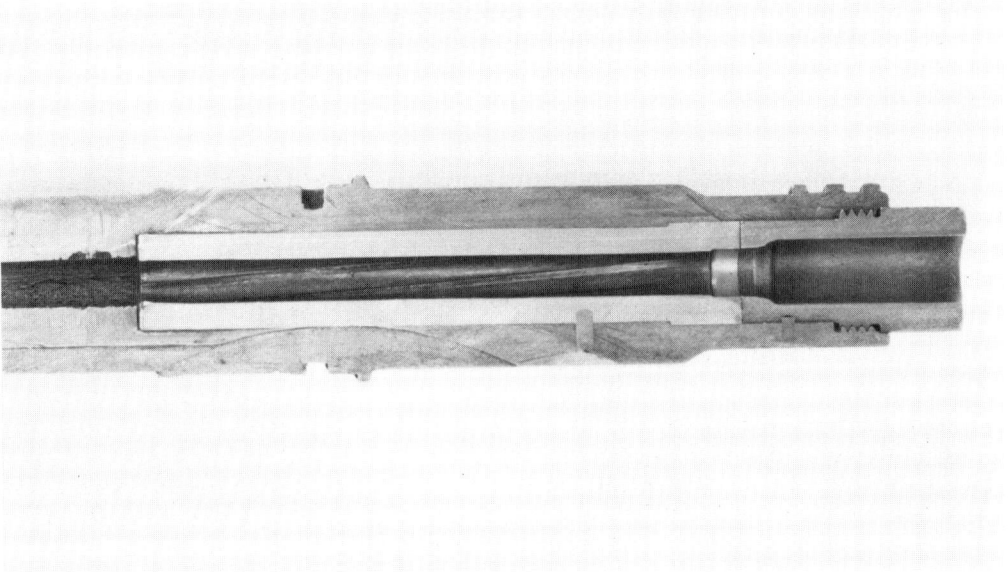

Fig 4.12 Typical Stellite Liner

One drawback to stellite liners is the difficulty of ensuring a perfect seal between the liner and the barrel. Rear obturation failures resulted in the rejection of their use by the UK.

Cooler Propellants

The amount of heat absorbed by the bore can be reduced by using either cooler or ablative propellants. Cooler propellants, as their name implies, have lower flame temperatures. However, for the same charge weight, cooler propellants produce lower muzzle velocities. If this is critical, it can be overcome by increasing the charge weight but this in turn increases the amount of heat produced by each round. In order to reduce bulk and weight high energy propellants are used in preference to cool ones. Ablative propellants can be used in the ammunition for large calibre weapons. This type of propellant contains compounds which smear the bore surface with a protective coat. Unfortunately this causes an increase in fouling and because of this they are not used by small arms at present.

Positioning of Hand Held Parts

It is generally accepted that $40^{o}C$ is the upper temperature limit for those parts of the weapon which are held by the firer. Such parts include the handguard for all weapons which are not fired from a bipod, the trigger, and the butt or cheek rest. All these areas become hot through conduction and in some cases radiation of heat from the barrel. Therefore their distance from the barrel affects the time taken for them to become uncomfortably hot. When it is not possible for reasons of design or cost to place these parts at a distance from the barrel then they must be insulated in some way.

The photograph at Fig. 4.13 shows the configuration of a conventional rifle, the US Armalite, and a buttless weapon, the French Famas rifle.

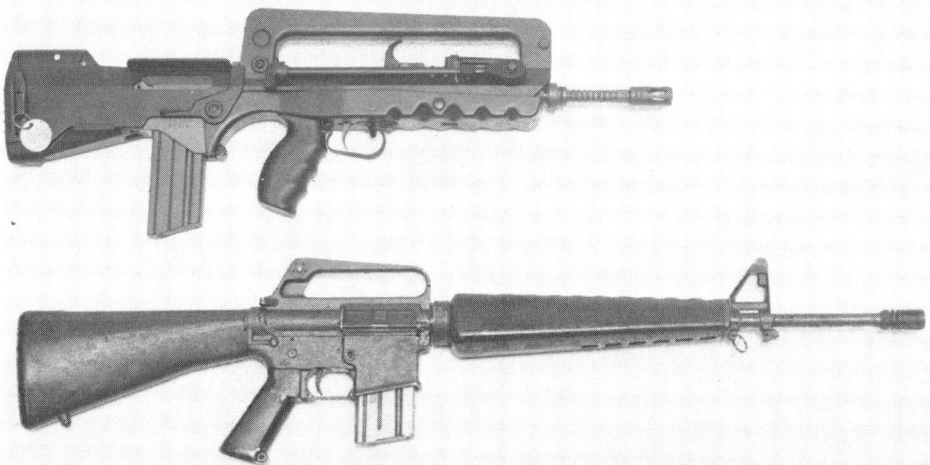

Fig 4.13 Configurations of the US Armalite Rifle and the French Famas Rifle

The handguards of both weapons are likely to overheat after the same number of
rounds have been fired. However the pistol grip and cheek rest of the Famas are
likely to heat more rapidly since they are both much closer to the barrel. Since
the handguard is the first part to become too hot to hold, the number of rounds
needed to make it reach 40°C becomes a limiting factor in the volume of fire
which the weapon can produce.

The photograph at Fig. 4.14 shows a soldier holding a UK L7A1 GPMG. As is
the case with all bipod mounted weapons it is unnecessary to hold the handguard.
Therefore these weapons can be fired for much longer than rifles before they
become uncomfortable to hold.

Fig 4.14 UK L7A1 GPMG (an adaption of the MAG58 GPMG)

THE EFFECT OF BATTLEFIELD REQUIREMENTS ON HEATING

Rifles

Rifles have traditionally been used to provide accurate single shot fire. To
achieve the necessary accuracy, rifles fire from closed bolts. The required
range has decreased from around 800m at the beginning of World War I to 400m
at the present. The longer range requirement is now being met by other weapons.

Approximately 40 rounds per minute is considered to be a reasonable rapid rate
of fire for aimed single shots. Under battle conditions a soldier might be expected to maintain this rate for about 2 minutes. The graph at Fig. 4.15 shows the
temperature reached under such conditions. There is no risk of cook off as the
chamber never gets hot enough to raise the propellant temperature above 200°C.

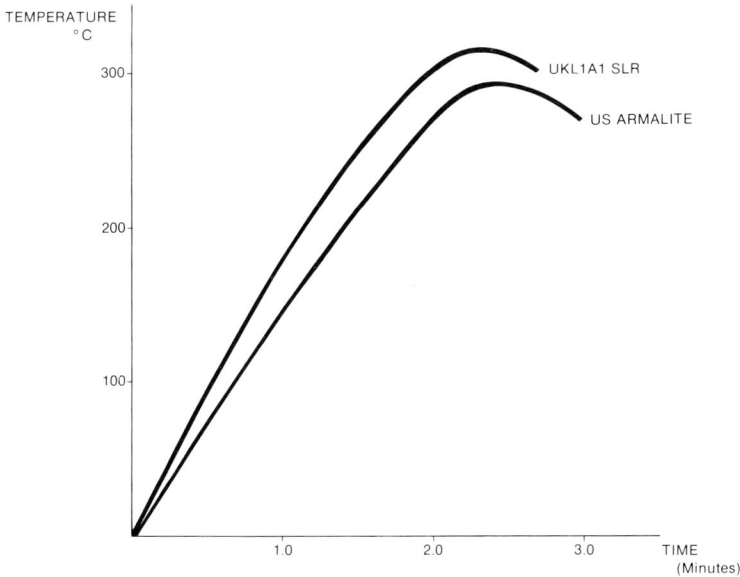

TEMPERATURE
°C

300

200

100

1.0   2.0   3.0

UKL1A1 SLR

US ARMALITE

TIME
(Minutes)

Fig 4.15 Barrel temperatures of rifles firing at 40 rounds per minute

Most modern rifles are capable of automatic fire. As the forward part of the barrel is unsupported the accuracy of all but the first round is poor. Well trained firers may be capable of keeping a short burst within a full sized target at 100m but most soldiers will be lucky to do this at 50m. Automatic fire is most effective at close range and it is for close quarter battles that assault rifles are fitted with an automatic firing facility. In these conditions rates of fire of up to 80 rounds per minute are easily achieved although the duration of the fire fight is unlikely to exceed 2 to 3 minutes at such rapid rates. Modern assault weapons may well run into cook off problems after about 200 rounds have been fired and it is likely that light weapons will have become too hot to hold with comfort. There is the possibility that this latter factor will limit the duration of fire so that cook off will not occur.

Section Machine Guns

Section machine guns are expected to provide accurate sustained volumes of fire. To be accurate out to the required ranges of 800-1000m the barrels have to be supported by bipods. As the chamber temperatures are likely to exceed the cook off safety limits even at modest rates of fire, machine guns fire from open bolt positions. The small loss of accuracy from the use of the open bolt is partially compensated for by the bipod support for the barrel. Any remaining inaccuracy is accepted in the interest of meeting the volume of fire requirements. Section machine guns are often used to engage targets whose exact location is not known.

It is therefore an advantage to have a spread of bullets, indeed it has been said in the past that some machine guns were too accurate.

The battlefield requirement for a section machine gun can be divided into three categories. Firstly the gun is used to hinder enemy movement and to discourage an enemy from interfering with the actions of friendly troops. This can be termed suppressive fire and is achieved by firing at about 50-60 rounds per minute within approximately 15m of the target. A section machine gun could be expected to sustain such a rate of fire for over 30 minutes. The graph at Fig. 4.16 shows the temperature rise for two different LMGs firing at such a rate.

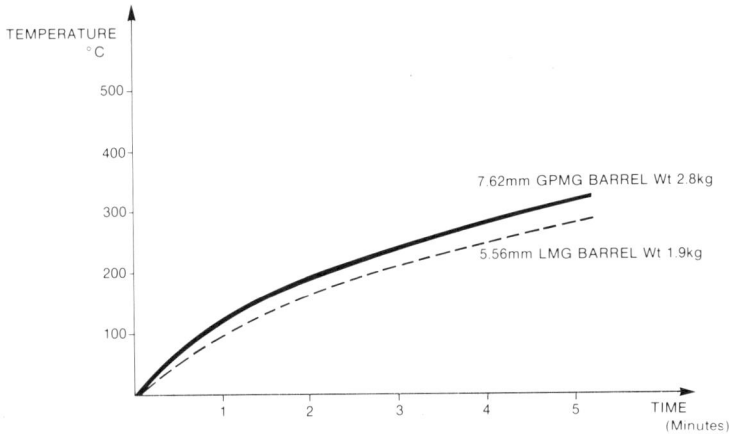

Fig 4.16 Temperature rise of typical LMGs producing suppressive fire

Secondly, section machine guns are used to prevent the enemy from returning fire or moving in any way. This effect is achieved when the rate of fire is raised to approximately 100 rounds per minute. In battle such a rapid rate could be expected to be kept up for perhaps 2 minutes. The graph at Fig. 4.17 shows the temperature rise under those conditions.

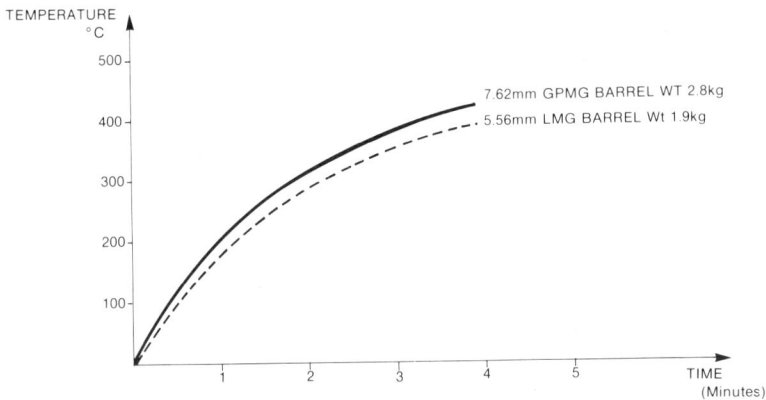

Fig 4.17 Temperature rise of typical LMGs firing at rapid rates

Thirdly, the section machine gun can be expected to fire for short intervals at the maximum practicable rate. For magazine fed weapons this is about 150 rounds per minute and could be higher for belt fed weapons. This intense rate is likely to be used on initial contact to gain ascendency over an enemy and is termed winning the fire fight. It is also likely to be used in supporting the final moments of an assault or in beating off a determined enemy attack. The graph at Fig. 4.18 shows the temperature rise of two LMGs firing at intense rates.

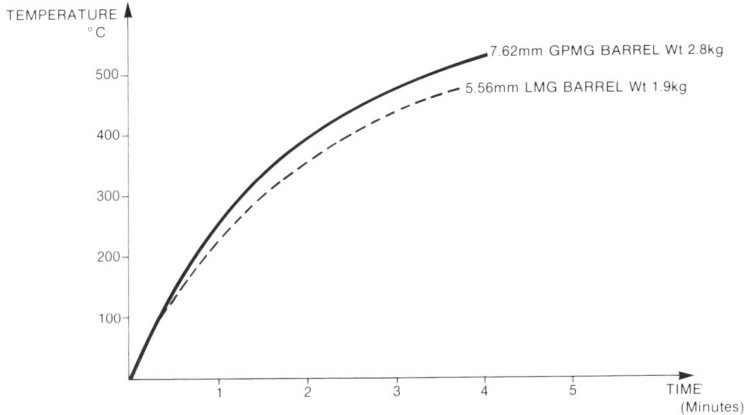

Fig 4.18 Temperature rise of typical LMGs firing at
intense rates

The overall requirement for a section LMG is that it should be able to sustain a rate of approximately 50 rounds per minute for up to 30 minutes. During this period the gun will be expected to fire at both rapid and intense rates for short intervals. The rapid and intense rates of fire will occur when the barrel is already hot. Such a requirement is a difficult one to meet for light weapons unless barrel changing is used.

Medium Machine Guns

The medium machine gun is expected to sustain a high volume of fire out to very long ranges. The Vickers, a classic example, could fire out to 4000m with the Mk VIII Z ammunition. Nowadays MMGs are only expected to fire out to around 2000m. To achieve a reasonably small spread the guns are mounted on tripods. They are therefore bulky and difficult to carry. Many modern machine guns have a dual purpose role as both the section and medium machine gun. These are called general purpose machine guns. The equipment which enables them to be converted from LMG to the MMG role is held in transport at company or battalion level. The photograph at Fig. 4.19 shows the UK L7A1 GPMG mounted on a tripod in its sustained fire or MMG role.

Fig 4.19 UK L7A1 GPMG (SF)

MMGs normally fire in continuous bursts of 25 rounds and therefore overheating is a major problem. These difficulties are overcome by heavy barrels which are often lined with some hard material such as stellite. These barrels are either changed every few hundred rounds or cooled by using water jackets.

Vehicle Mounted Machine Guns

Vehicle mounted machine guns can be fired in bursts of up to 25 rounds but are quite often fired in much shorter bursts. As weapon weight is less of a problem most VMMGs have heavy lined barrels which can be changed. Figure 4.20 shows the .30 Browning which has a slow cyclic rate of fire of 400-550 rounds per minute and a heavy barrel weighing 3.3 kg. These two factors help to reduce the need to change barrels often.

Fig 4.20 0.30 Browning VMMG

SUMMARY

The heat produced when a small arm is fired is absorbed by the barrel faster
than it can be dissipated to the surroundings. Therefore as more rounds are
fired the barrel gets progressively hotter. This heat is conducted to other parts
of the weapon which can become uncomfortably hot. Good design should prevent
this problem from occuring under normal battlefield conditions. The heat conducted to the chamber is in turn absorbed by an unfired round in the chamber. If
the temperature of the propellant is raised above 200°C it will ignite spon-
taneously. This process of cook off can be a problem with rifles which fire from
closed bolt. However it is only likely to arise after prolonged rapid firing and
so the risk is considered acceptable. The chambers of section machine guns
readily exceed the cook off temperature safety limits and so these weapons are
designed to fire from an open bolt. When bore temperatures approach 450°C-
500°C wear and erosion are likely to occur suddenly causing extensive damage
to the bore which leads to inaccurate fire. The problem can be solved to some
extent by the use of thicker barrels or special bore liners and can be prevented by
changing the barrel before the critical temperature is reached.

SELF TEST QUESTIONS

QUESTION 1   Why do small arms overheat?

Answer   ....

....

QUESTION 2   What are the three main problems caused by overheating?

Answer   ....

....

QUESTION 3   Why is radiation so effective for heat transfer at high
temperatures?

Answer   ....

....

QUESTION 4   Why is it necessary to make a small arms barrel a heat
reservoir? How is this achieved?

Answer   ....

....

QUESTION 5   Will a round cook off if the outer barrel surface temperature
is 230°C?

Answer   ....

....

QUESTION 6   What is the difference between wear and erosion?

Answer   ....

....

QUESTION 7   What is gas wash erosion?

Answer   ....

....

QUESTION 8   Under what conditions will fins on a barrel be an effective
method of losing heat?

Answer   ....

. . . .

. . . .

QUESTION 9   What is the most effective method of avoiding overheating
               problems for an LMG?

               Answer   . . . .

QUESTION 10   What do you understand by the term "suppressive fire"?

               Answer   . . . .

. . . .

ANSWERS ON PAGE 182

5.

Chance of Hit

Why Accuracy

Although the need for great accuracy is sometimes questioned, the traditional UK attitude is that each shot must count. Perhaps this view was highlighted by experience in the Boer War when the marksmanship of the Boers proved so effective. To achieve a high standard of single shot accuracy out to long range requires weapons and ammunition to be well designed and engineered. Much time and ammunition has also to be used to train the firer. This can be difficult because the modern soldier must become skilled in many aspects other than musketry.

The photograph at Fig. 5.1 shows a soldier carrying the UK L1A1 SLR.

Fig 5.1 On patrol in Northern Ireland

This weapon is an adaption of the Belgian FN rifle; one modification being the removal of the automatic fire facility. It has a ring aperture backsight and can be fitted with an optical sight. The weapon has been made to optimise the firer's chances of hitting a relatively stationary target. If a target takes violent evasive action, particularly at short range, then the hit chance can be improved by firing a burst into the target area. The ability to acquire rapidly a fleeting target is assisted by using a weapon fitted with an open (V) backsight. All Soviet assault rifles are fitted with open sights and can fire bursts. These two examples indicate that the accuracy requirement of small arms is open to different inter-pretation. Some of the factors which should be considered by the soldier are covered below.

The requirement for great accuracy of each shot could depend on the following factors. Firstly the target may be stationary or moving slowly, so a single shot could have a high hit chance. Such a target is often engaged by snipers out to 800m. Secondly the target could be surrounded by people whom the firer did not wish to injure: terrorist situations are an example. Thirdly the firer's morale is maintained if he is confident that his weapon can fire accurately.

On the other hand the army's tactics may not envisage personal weapons being used beyond about 300m. The need to engage targets with a high degree of accuracy beyond that range can be done either by specialists armed with small arms or by other weapons. In short range engagements, the volume of suppres-sive fire is most important. This can be best achieved with automatic weapons. Keeping an enemy's head down can be done by a near miss as well as by a hit, so pin point accuracy is not so important.

All these factors have to be balanced so that the designer is given a clear indica-tion of the accuracy needed of a weapon. The details of how small arms fire is made to be accurate is covered in the rest of this chapter.

Definitions

It is important at the outset to understand the difference between two charac-teristics:

Accuracy is an ability to apply rounds to a point of aim.

Consistency is an ability to prevent successive rounds from spreading out in a dispersed pattern.

To clarify these definitions requires an understanding of the meaning of the term Mean Point of Impact (MPI). As no two rounds hit the same point (except on the rarest of occasions) the pattern formed by any number fired into a target with the same point of aim is called a group. The MPI is the geometrical centre of the group. The group will have a horizontal and a vertical spread. The MPI can be calculated accurately from the horizontal and vertical dispersions by simple mathematics. In fact a surprisingly close approximation to the MPI of

most groups can be arrived at by visual estimation.   Take two groups of 5 rounds
on the targets illustrated below.

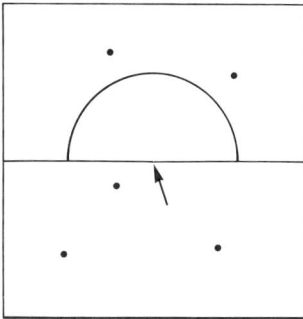

Accuracy shooting   Consistency shooting

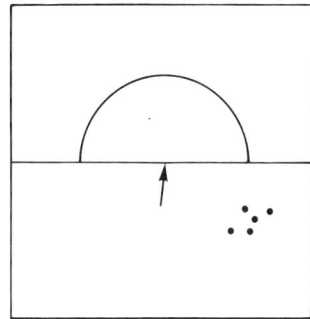

Fig 5.2 Comparison of Accuracy and Consistency

The aiming point in both cases is at the centre of the target (arrowed), In the
left hand diagram the MPI cannot be far away from the centre, so the accuracy
is quite good.   The dispersion is considerable and consistency therefore is poor,
and some component in the system is not playing its part well.   In the right hand
diagram the MPI is obviously low right, and accuracy is poor, but the consistency
is good, with a nice tight group.   Obviously the soldier is looking for a weapon
system which gives good consistency and a means of applying the tight groups
which can be obtained to the aiming point to give accuracy.   By dint of careful
design, quality manufacture and good handling, this can be achieved.

Figure of Merit

Before looking at the components there is one more digression to see how ammunition is judged.   In the UK this is by the Figure of Merit (FOM) known in
other countries as the Mean Radius or Mean Radial Deviation.   It is derived as
follows.   A number of rounds is fired and the MPI calculated.   This is done by
taking a datum point, usually the bottom left round, or the corner of the target,
and measuring the vertical and horizontal displacement for each round.   The
average of the displacements in each plane gives the MPI.

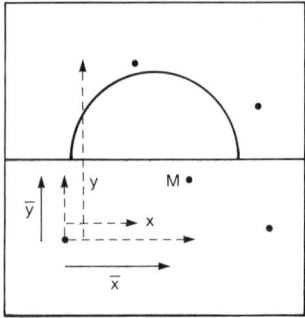

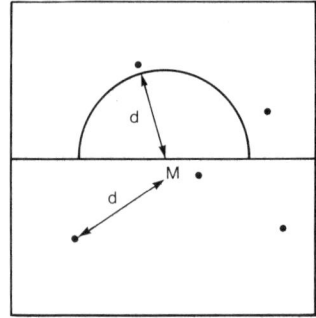

MPI

Measure x for each shot
Measure y for each shot
Find the means $\bar{x} = \langle \dfrac{x}{n}$

$\bar{y} = \langle \dfrac{y}{n}$

MPI is at M = $\bar{x}$, $\bar{y}$

FOM

Measure d for each shot
FOM = $\langle \dfrac{d}{n}$

Fig.5.3 Method of Measuring MPI and Figure of Merit

$\bar{x}$ is the average of the horizontal displacements

$\bar{y}$ is the average of the vertical displacements

The MPI is $\bar{x}$ horizontally and $\bar{y}$ vertically from the datum

For those who like mathematical notation the MPI is $\bar{x}$, $\bar{y}$ where

$\bar{x} = \dfrac{x}{n}$ and $\bar{y} = \dfrac{y}{n}$ (being the sum of, and n the number of shots)

The direct distance of each round to the MPI is measured, and the FOM is the average of this measure of dispersion. Shown mathematically

$$\text{FOM} = \frac{d}{n}$$

In obtaining the FOM an attempt is made to eliminate the errors which could be introduced by factors other than the component whose FOM is being measured. For example, if the FOM for ammunition is required, a heavy fixed barrel is used in an enclosed range free from the worst effects of wind and temperature fluctuations. Inevitably there will be some outside influences, but with the bulk of them removed the FOM is the most realistic for the ammunition lot being tested.

THE AMMUNITION

The technical specification for a round of small arms ammunition is distinctly formulated. Dimensions are held to close tolerances. Cartridge cases are filled with considerable care to get the charge weight correct. Modern ammunition production processes are very highly automated, and at frequent stages automatic checks are made on dimensions and weight. Despite the enormous number of rounds coming off the production lines every one has been subjected to ingenious measuring devices which automatically reject any round that is not within the tolerances. When a Lot (usually 10,000 rounds of small arms ammunition) is produced, it is subjected to proof. This is a process conducted by quality assurance personnel who not only weigh and measure a proportion of the rounds but also fire a number to see that they perform within Acceptable Quality Levels. A part of this process is to obtain for each Lot the FOM. For UK service use the acceptable FOM for 7.62 mm L2A2 ball ammunition is 8 inches at 500 yards. The best that the Royal Ordnance Factory at Radway Green has produced is 2.8 inches. As the FOM is an average, some rounds in a FOM group must fall outside the FOM figure. Simple statistics will show mathematically that 99.9% of rounds fired will hit an area 3 times the FOM. The component of consistency, and therefore accuracy, which is contributed by the ammunition can thus be seen to be quite good. No in-service 7.62 mm ammunition should be responsible for a dispersion greater than 24 inches at 500 yards, and it is frequently much better.

THE WEAPON

Adjustment of the sights can move the MPI of a group to the point of aim. This is the purpose of zeroing. The way in which sights are mounted on a weapon, and idiosyncrasies of hold and aim by the firer, may mean that no matter how small the group, it does not hit the point of aim. So a series of groups are fired and the sight arrangements adjusted until this is achieved with that weapon for that firer from that fire position. Assuming this is to be done, and assuming that the potential for dispersion in the ammunition has been reduced to the minimum, it is now necessary to ensure that the weapon does not introduce a dispersion that swamps that of the rounds. Once again we are looking for consistency.

Mechanism

The accurate dimensions and minimal round to round size variations of the ammunition helps to keep the fit in the chamber similar with every round. When the bullet is fired the intention is to maintain a chamber pressure of least possible variation. If the space in the chamber of a weapon varies between rounds, so will the chamber pressure. The weapon mechanism is designed both to contain this pressure and to give it little scope for round to round change. At the moment of firing the bolt will either be locked tightly to the body, or with a blowback weapon like the West German G3 rifle, the delay rollers will have a similar effect. All this is done to reduce chamber volume variation. Additionally the Cartridge Head Space (CHS) is carefully controlled. This is a measure of the gap between the cartridge and the bolt face which will be described in more detail in Chapter 8. Suffice here to say that it is a parameter designed to vary as little as possible.

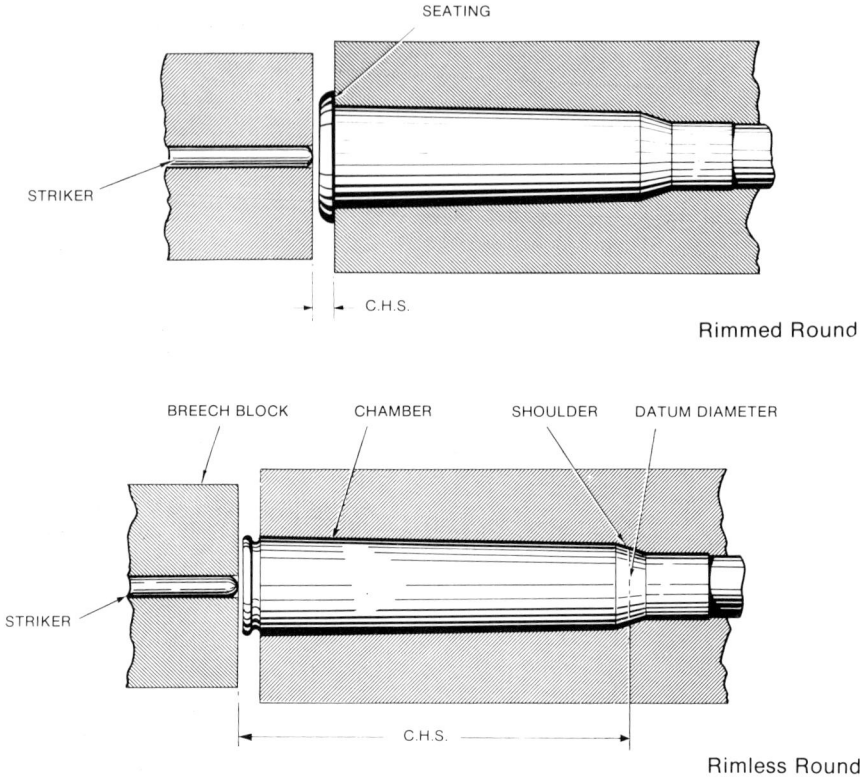

Fig 5.4 Cartridge Head Space

An important element in the mechanism affecting accuracy is lock time. This is the time from squeezing the trigger to the cap or primer in the cartridge case being fired. The time should be as short as possible since the correct aim was being taken when the trigger was pulled, and it may be lost if the lock time delays cartridge initiation. The majority of rifles and pistols fire from a closed bolt and lock time is therefore short, in the region of 7 milliseconds. Even so it can be seen how influential this time can be if compared with the time the bullet is in the bore of about 1.3 milliseconds. A specially designed and prepared competition rifle may have a lock time in the region of 1 or 2 milliseconds. Automatic weapons need usually to fire from an open breech so that the heat can be dissipated between bursts. In this case the working parts have to travel forward before the firing pin strikes and lock time is rather longer, typically in the order of 25 milliseconds.

The Barrel

Ammunition is manufactured to close tolerances so that it will fit well into any weapon for which it is designed. The chamber and barrel of the weapon have to be

made to similar accuracy in order that round to round and weapon to weapon variations are minimal.   The fit of any round in any chamber and its passage up the
barrel should be as closely repeated as possible.   Needless to say, in an imperfect world there are various causes of imperfect repetition, quite apart from the
impossibility of manufacturing to perfect dimensions.   There are problems of
barrel wear, vibration and droop.

Barrel Wear

The problems of barrel wear were discussed in Chapters 2 and 4.   The extreme
occurence when a barrel is 'shot out', or so worn that catastrophic dispersion
ensues, is relatively rare because designers take steps to reduce the likelihood
of it happening.   A steady wear of the inside barrel surface is inevitable, however, and the way in which it is overcome is the mechanism called set-up.   The
rear end of a bullet is not jacketted as are the sides.   When the gases from the
cartridge case force the bullet into the grooves of the rifling, all the force is
against this open end of the bullet.   The effect is to expand the base of the bullet
in such a way that a bulging rear end fits tightly into the rifling.   The minute
changes in dimension of the barrel due to wear are automatically compensated for
by this set-up action.   The result is that the all important parameter of pressure
is kept very nearly constant, round to round, and another potential cause of dispersion is eliminated.

Vibration

The stresses and strains involved in forcing a bullet up a barrel will produce
vibrations.   The way in which the barrel is held by both the firer and the furniture results in most of these vibrations being in the vertical plane.   Obviously, if
a barrel has vibrated or jumped upwards at the moment of bullet exit, the trajectory will be higher, and the point of impact at variance with the point of aim taken
when the trigger was pulled and the cap initiated.

To avoid errors which are induced by barrel vibrations, the designer controls the
length of the barrel which is free to vibrate so as to ensure that all bullets leave
the muzzle as it moves upwards.   If a bullet has a higher muzzle velocity it will
leave the barrel before the jump is at its highest.   A slower bullet, leaving later,
will be where the jump is a maximum.   But as the muzzle velocities result in a
flat trajectory when fast, and a high trajectory when slow, there will be a cross-
over point, as shown in the diagram at Fig. 5. 5.

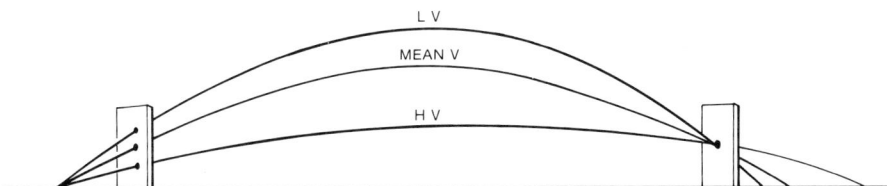

Fig 5.5 Compensating Range

From one weapon all bullets, which have muzzle velocities within agreed tolerances, should theoretically go through the same hole at one particular range.
That range is referred to as the compensating range for that weapon. The compensating range should ideally be the battle range of a weapon.

For weapons firing burst, designers select rates of fire which will either fit in
with the vibrations to minimise their effect, or will so slow the rate of fire, as
with the single shot UK Rarden cannon, that the vibrations have died before the
next round is fired. The Americans are so impressed by the resulting accuracy
of Rarden that they are using it in research, when the firing of rounds is elec-
tronically controlled to ensure that it happens only when vibrations are at a minimum. Vastly improved measures of accuracy have been achieved.

The sights on a weapon are generally calibrated for the compensating range, but
it is as well to note that there are other variables which the designer must take
into account. One of these is drift. The spin imparted by the rifling is an essential factor in accuracy because it stabilises what might otherwise be a wobbly
projectile. A side effect of the spin is that the air friction involved turns the bullet very slightly, and this is known as drift. It is of a very small order, but has
to be taken note of, and is enhanced when the wind blows across the line of flight
in the same direction as the drift. The other variable is termed yaw. This
happens because a bullet on exit may have a residual contact with the barrel on
one side, thus tending to distort its straight flight. This will temporarily be
magnified because the bullet surface will not have been presented symmetrically
to the aerodynamic effect of air friction. So for a certain distance the bullet will
have a vortex or spiralling action known as yaw. After a few metres the spin
from the rifling stabilises the yaw, but it is unwise to measure accuracy on anything shorter than a 25 metre range, because up to that distance the yaw may
introduce inconsistency. Shot out barrels cannot provide sufficient spin to overcome yaw. The photograph at Fig. 5.6 shows the yaw on a bullet just after it has
left the muzzle.

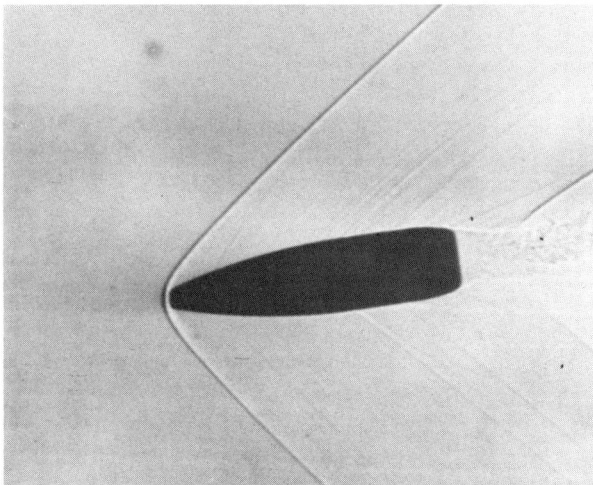

**Fig 5.6 Yaw in a 7.62mm bullet shortly
after muzzle exit**

Droop

Droop is a comic sounding effect which will take place in any barrel which becomes heated by firing. The heat effectively softens the metal and the barrel quite literally droops in consequence. The longer barrels of tank and field guns are more susceptible than small arms to this effect, but although the extent is small it is enough to cause the MPI of a weapon to drop as the following table shows.

TABLE 5.1   Effect on MPI of L7A2 GPMG Barrel when Heated

Range 200 metres	
Temperature °C	Drop of MPI mm
100	75
300	127
500	406
600	610

A method of counteracting droop is to support the barrel well forward as in the US M60 machine gun which is shown in Fig. 3.5.

ADDITIONAL DESIGN FEATURES

Quite clearly it is the designer's task to incorporate in his weapon and ammunition as many of the features which lead to consistency as he is able. This inevitably leads to certain elements of the design being in conflict with others, and a compromise has to be struck. Some of the most fabled accurate shooting was from the weapons used by hunters of buffalo. They achieved this accuracy in part by having very long, heavy barrels. Even today .22 competition rifles will be found to have apparently unnecessarily heavy barrels, but this reduces vibration almost entirely. The buffalo hunter's bullet was heavy too, and it was propelled by a large amount of propellant which built up a very high pressure in the chamber giving the bullet a very high velocity. This meant that it travelled in a flat trajectory closer to the sight line (a straight line) than a slower bullet. The latter would have had a longer time to feel the effect of gravity and would fall in an arc. A heavy bullet has more inertia and is less likely to be affected by side winds, or other causes of deflection like leaves or rain.

On a long barrel can be mounted a backsight and a foresight at some distance from one another; this is termed a long sight base. This long sight base reduces the potential for sighting error which can creep in with a short sight base. As the diagram at Fig. 5.7 shows, misalignment of the sights leads to a greater angle of error with a short sight base.

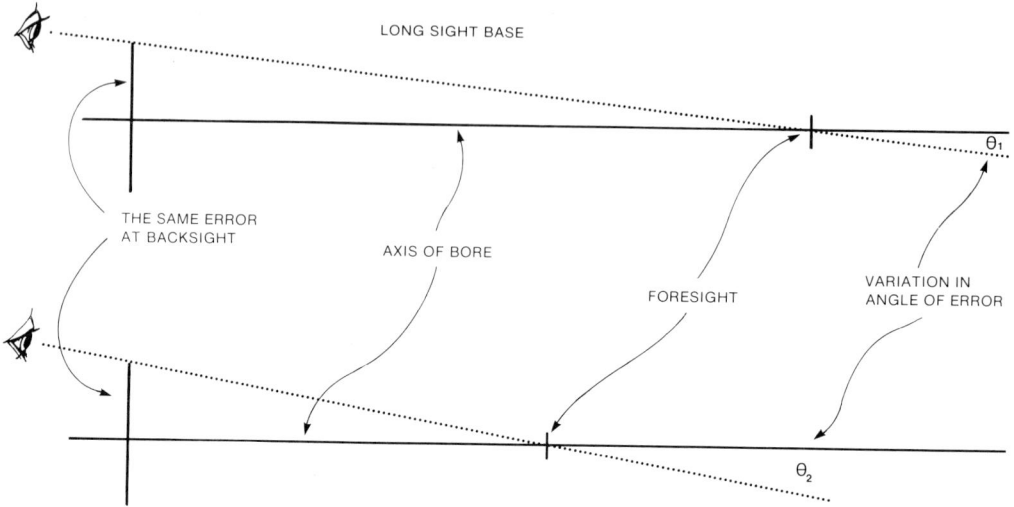

LONG SIGHT BASE

θ_1

THE SAME ERROR
AT BACKSIGHT

AXIS OF BORE

FORESIGHT

VARIATION IN
ANGLE OF ERROR

θ_2

Fig 5.7 Short Sight Base

All of the above was necessary to ensure a hit and a kill of a large animal with one shot by a well placed bullet at long range. But the weight of the weapon and ammunition was very high. The recoil was tremendous and required a very strong man to withstand it and not be put off his aim by the expectation of the kick. Aids to holding were used while aiming, such as slings and even monopods.

A parallel can be seen in the requirements for the military sniper, who will often be armed with a specialist rifle and trained to use aids to long range accuracy such as slings. Referring back to the chapter on requirements, it will be re-called that the average soldier does not want to be laden with a heavy, hard to control weapon, and so the designer seeks to achieve accuracy in other ways. For the reasons described in Chapters 1 and 3 the most common small arm is going to be a light, short weapon of small calibre. Many of the components con-tributing to consistency have been diluted, but there are still a few tricks up the designer's sleeve. He ensures a high muzzle velocity with a considerable charge weight and a propellant of high impulse characteristics. A high rate of spin and elongated bullet shape achieve a stable flight in still air. The handling charac-teristics are such that in the most inexperienced of hands it is still possible to apply fire accurately; and the design and fitting of the sights is calculated to be simple but effective.

Small arms are frequently so designed that there is a turning moment introduced on firing when the recoil acts through the butt into the firer's shoulder. A turning moment is the technical term for an inclination of an object to turn about a pivot. One turning moment in a small arm's action is caused by the centre of gravity of the weapon not being on the bore of the barrel as shown in Fig. 5.8. With the centre of gravity offset a distance d and a recoil force P there will be a turning moment of force Pxd.

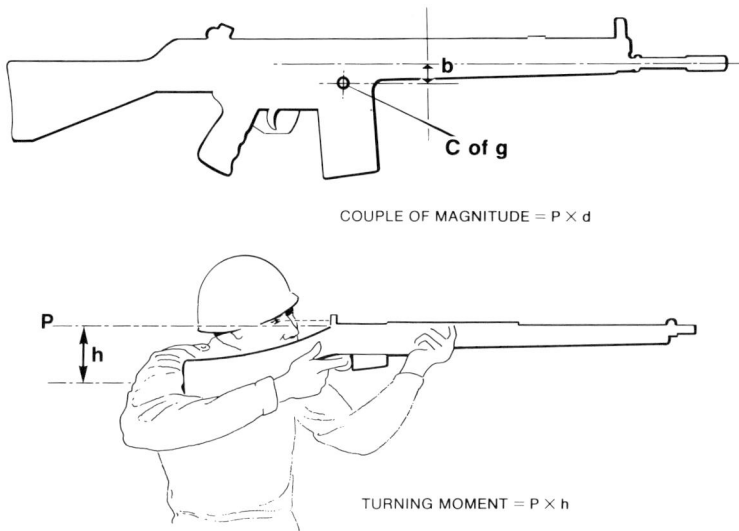

COUPLE OF MAGNITUDE = P × d

TURNING MOMENT = P × h

Fig 5.8 Factors Affecting Muzzle Movement

An easier turning moment to appreciate is that brought about because a rifle butt is angled from the line of the barrel. The point where it is cushioned by the firer's shoulder is offset a distance h from the axis of the bore, and a recoil force of P will cause a turning moment Pxh. The result of this can often be seen in the way a rifle jumps in the hands of an inexperienced firer, and is even more obvious when a pistol is seen fired.

The better firers learn to control this tendency for their shots to go high, or at least repeat the error with some consistency on single shots, but it is very difficult indeed to control with automatic fire. It is a very common fault seen in burst fire for the shots to climb towards or above the top of the target. If the weapon design is such that the recoil force is directed horizontally back into the shoulder to eliminate the turning moment, the sight has to be mounted very high above the barrel to allow the firer to get his eye to it. The Bullpup design of rifle, of which the UK Individual Weapon is an example, has this feature of a straight through butt, and no turning moment in consequence, but it, like the Armalite, has to have its sight mounted high. There are other ways of trying to reduce the turning moment effect, one of these being the compensator, which is described in Chapter 6. Another is so to distribute the weight of components like the magazine, that the centre of gravity does lie on the axis of the bore.

The average soldier is deterred from holding his weapon properly and taking a careful aim if the weapon is heavy and has a high recoil. The sheer weight is tiring, both to carry into battle and also to hold in the aim. A recoil which bruises the shoulder or cheekbone will make the soldier flinch in anticipation at the very moment when his hold and aim should be most steady. A loud report is likely to make him close his eyes, rather than keep a correct aim picture throughout trigger operation, lock time and bullet exit. The lighter, small calibre, modern weapons should, hopefully, be far less tiring for the soldier to carry, and the lesser recoil and noise will contribute significantly to better results for the average or poor shot. The better shots and dedicated marksmen are unlikely to improve their standards with these weapons, but they are in the minority, and it is more often the average or poor shot who comes up against the enemy or terrorist, and whose shot has to count.

SIGHTS

An essential part of a small arm system, if it is to have a reasonable chance of a hit, is the sights. Simply pointing the barrel as is done with much shot or scatter gun shooting, is not good enough when there is no spread of shot. An accurate alignment of the barrel with the line of sight to the target has to be achieved. For small arms this is done with the simple direct sighting systems. More complicated indirect sights for directing the fire of guns, mortars and some machine guns are not considered in this chapter.

Light travels in straight lines whereas a bullet trajectory curves due to gravity, and sometimes due to wind deflection. Careful note of these deflections must be taken by the firer and adjustments made to the sights either as a range elevation or as an aim off for wind. Apart from this the purpose of the sights is to present to the firer's eye as clear a picture as possible of the target under all conditions, and in so doing to direct the line of fire of the weapon to the target. At the same time it should introduce the least opportunity for error. There are 4 main types of sight - open, aperture, optical and lensatic.

Open sights

Open sights are cheap and simple. They usually rely on a V backsight and a post foresight which have to be aligned on the target as shown in the diagram at Fig. 5.9 opposite.

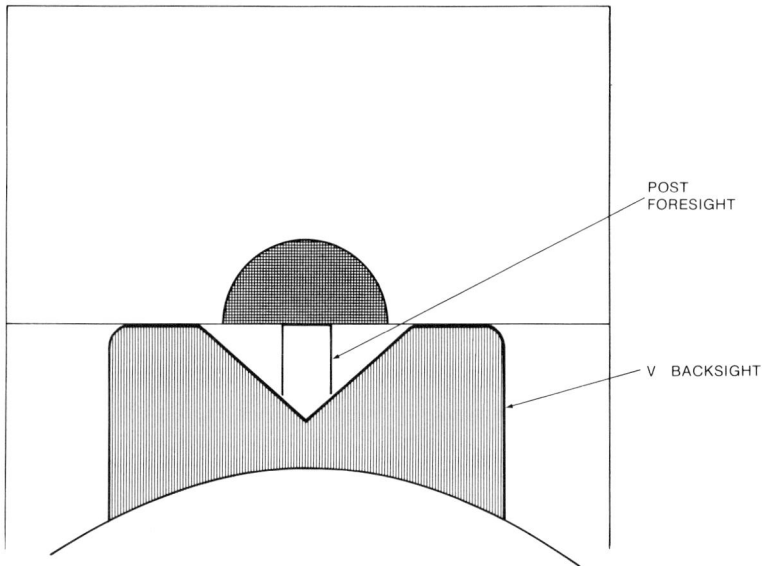

Fig 5.9 The Open Sight

They are prone to error because the post may not be central in the V, it may be above or below the top line of the backsight, and it may not be brought fully up to the point of aim on the target. All of these can introduce error, and more so if they are on a short sight base as these sights tend to be - on a SMG or assault rifle.

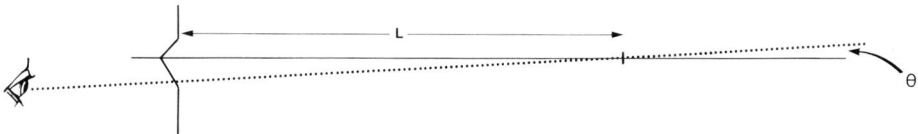

Fig 5.10 Error in an open sight

With the sight line off centre in the V, the shorter the sight base L the greater the potential angle of error θ. The great advantage of the open sight is that it can be brought into the aim more quickly than other types of sight, since there is no interference with observation of the target whilst the weapon is raised to eye level.

Aperture Sight

The aperture sight cannot be quite so quick to use because of the material round the aperture which obscures the target momentarily whilst the weapon is raised into the aim. However, the aperture tends automatically to centre the sight picture and reduce the possible errors described with the open sight. Also if the aperture is kept small then any error in alignment should be small too.

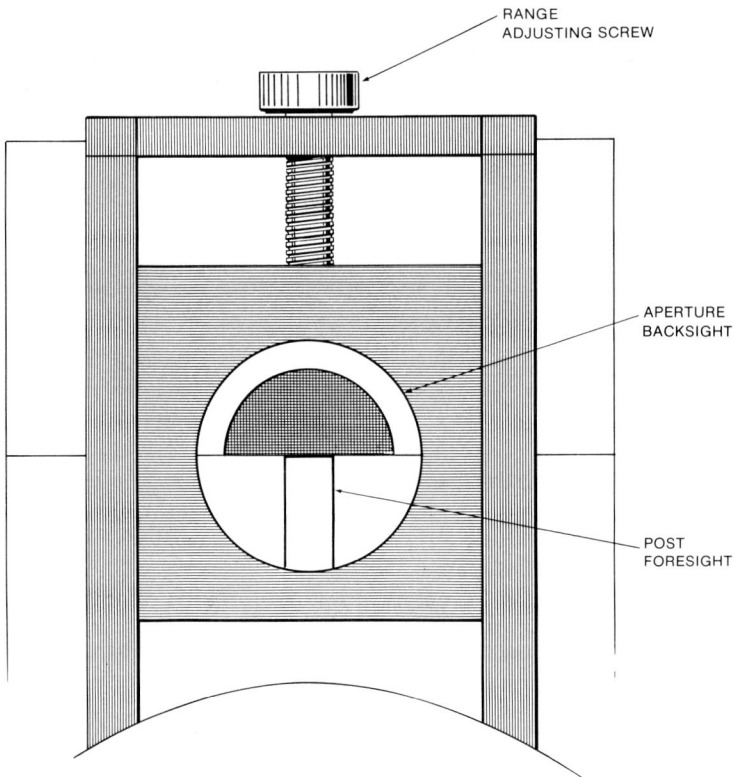

RANGE
ADJUSTING SCREW

APERTURE
BACKSIGHT

POST
FORESIGHT

Fig 5.11 The Aperture Sight

Sighting Errors

The human eye has difficulty in seeing simultaneously and clearly objects which
are at greatly different distances from the eye.   The eye cannot focus on both the
target and the foresight and backsight of an open sight.   The firer must quickly
adjust his focus from the target and foresight onto the backsight.   At the moment
of firing one of these groups will be out of focus and so the shot must be
inherently inaccurate.   The problem is overcome to some extent with the aperture sight as there is no need to focus on the backsight; the eye theoretically
looks through the centre of the ring.

Under low light conditions the depth of focus of the eye deteriorates as the pupil
expands to let more light onto the retina.   The open sight is very difficult to use
but the aperture sight's performance can be maintained in poor light conditions
by varying the size of the aperture to match the diameter of the pupil.   Even so
shooting in poor light is usually more effective if the weapon mounts an optical
sight.

Optical Sights

In an optical sight the lenses present the target and the aiming mark (the graticule) in the same plane to the eye.   This ability in effect to focus foresight, backsight and target all at once brings a consequent reduction in the potential errors.
In addition the optical sight may offer some magnification (the telescopic sight)
which allows small and indistinct targets, and particularly those in poor light, to
be engaged more effectively.   The optical sight is thus able quite substantially to
extend the range at which a target can be engaged, and its universal use by
snipers is a result of this characteristic.

**Fig 5.12 Optical (telescopic) sight on 7.62mm
L42 A1 rifle**

There are drawbacks to optical sights. They are generally more vulnerable to damage and to condensation or misting up. They are likely to be weighty and expensive. Magnification can make acquisition of the target more difficult since firstly the field of view through the sight becomes smaller as the magnification increases; secondly the eye can take a few moments to adjust to the magnified picture and thirdly movement of the sight produces an exaggerated movement of the picture. A sight that can avoid most of these drawbacks but help to reduce firer error is one that has unity magnification or very near to it. An early example of a unity magnification sight was on the UK experimental EM2 in the 1950s. A good modern example is the 1.5 magnification on the Austrian Steyr.

Lensatic Sights

The lensatic or collimator sight places a translucent graticule in the focal plane of the eyepiece. A simple convergent lens is so placed that emergent light rays are parallel to each other and also to the axis of the sight, and the bore of the weapon. If the target coincides with the graticule, the axis of the sight and hence the bore, must be directed at the target.

A development of this system is the Singlepoint sight which is shown below in Fig. 5.13.

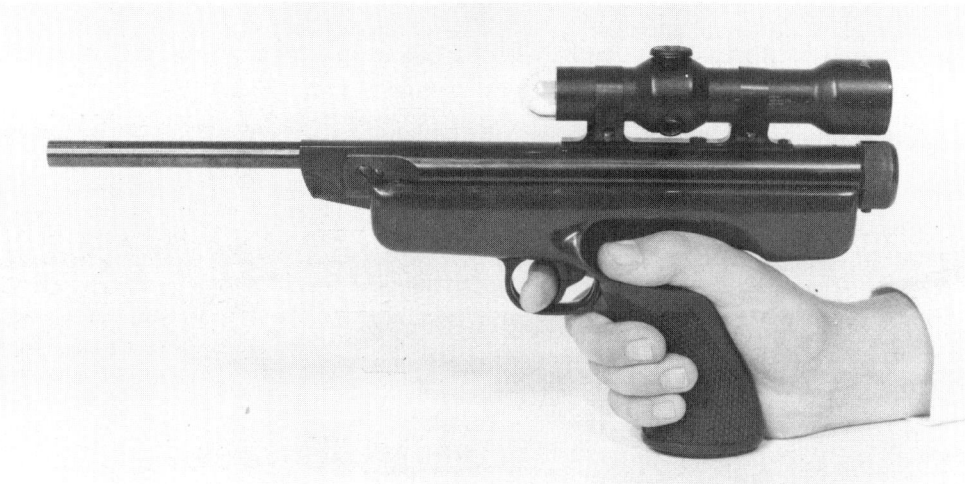

Fig 5.13 Singlepoint sight on an air pistol

In a Singlepoint sight a Beta light source is placed forward of a pinhole. The convergent lens is used to make the rays of light from it parallel to the axis, and this is presented to the firer's eye as a coloured spot of light. The firer keeps both eyes open, looking at the target with the free eye. If his eyeballs are coordinating properly this will result in the axis of the sight being pointed at the target. Unfortunately about 1 in 10 people do not coordinate properly and cannot use a Singlepoint. The sight can have advantages for target acquisition and the quick snap shot, but it is less suitable for carefully aimed shooting and unlike an optical sight does not improve target acquisition in low light conditions.

Mounting

Whatever sight is selected for a small arm its mounting is extremely important. The problems likely to arise from a short sight base have already been mentioned. To complicate this, the position in relation to the firer's eye is critical. It must be neither too high nor too low, too well forward or to the rear, and varies with the type of sight. The eye can be pressed close to an optical sight but must stand off from a backsight. A comfortable position for the head and perhaps a place for the cheek to rest can do much to steady a firer and to ensure that he takes a good aim. The mounting must be reasonably rigid to withstand handling and the considerable acceleration or G forces which are experienced on firing. The location of any sight which is dismountable must be to close limits of tolerance otherwise it would be improperly aligned, or off zero, when it was replaced. The mounting or the sight itself must incorporate a facility for zeroing. Zeroing was described briefly earlier, and is the process whereby the firer adjusts the sights to move the MPI of a group to the place he aimed at. In other words, to improve the accuracy of his shooting.

The Chance of a Hit

Now that we understand consistency, accuracy, and the ways in which ammunition, weapon, sights and firer all contribute, how do we assess the chance of a hit?

We saw that the Figure of Merit can tell us in mathematical terms what the ammunition can achieve, and also the weapon. This can simply be turned into a probability figure - that is a 0.6 or six tenths or 60% chance of hitting a target of a certain size at a certain range, knowing the FOM. The difficult figure to arrive at is the chance or probability introduced by the firer. Unfortunately the chance of a hit for the combined components - the weapon system - is the product of their individual chances. Should the ammunition be 0.9 and the weapon be 0.8 for a certain target size and range, their combined chance of hitting that target is 0.72. Generally it is the firer who introduces the largest error, or who contributes the least probability figure. If we introduce his contribution at the interface with the weapon system shown above of probability 0.72, and his chance of hitting is 0.6, then the total probability of that target being hit is 0.432. The only way to derive the firer's contribution is empirically. This is to say that, knowing the contribution of the weapon and ammunition, the actual chance of a hit, when the firer is included, is discovered from firing. It is then a relatively simple matter to calculate what probability the firer introduced. The manufacturer produces weapons and ammunition with care to enable them to afford a good chance of a hit. It is the soldier who must play his part by training, care and confidence in the weapon system, not to introduce errors which swamp those of

the weapon components.   The pistol is the ultimate example of this problem.   The
weapon system, rigidly held in test conditions, typically produces a standard
deviation of less than 1 mil.   (1 mil being the angle subtended by an arc of 1m at
1000m).   An average shot will introduce errors up to 40 mils.   The only way to
improve on the pistol's proverbial chance of hitting the barn door is for the
firer to train to handle it properly.

SELF TEST QUESTIONS

QUESTION 1   Briefly describe Figure of Merit and state the acceptable FOM
              for UK 7.62 x 51 mm ball.

              Answer   ....

                        ....

QUESTION 2   What is known to increase barrel wear?

              Answer   ....

QUESTION 3   What is droop and its effect?

              Answer   ....

                        ....

QUESTION 4   What advantage has a long sight base over a shorter one?

              Answer   ....

QUESTION 5   Round to round variations of muzzle velocity should be kept
              to a minimum.   Why is this so and how is it achieved?

              Answer   ....

QUESTION 6   What effect on trajectories have Muzzle Velocities (MVs) either
              side of the mean?

              Answer   ....

                        ....

QUESTION 7   Name the three elements in a weapon system which fundamen-
              tally affect hit chance.

              Answer   ....

                        ....

                        ....

QUESTION 8   What two causes of turning moments in a conventional rifle can
              contribute to inaccuracy?

              Answer   ....

                        ....

QUESTION 9   What must a sight mounting ensure?

Answer   . . . .

. . . .

QUESTION 10   Give some reasons why accuracy is still an essential in a
small arm.

Answer   . . . .

. . . .

ANSWERS ON PAGE 183

6.

Muzzle Attachments

Muzzle attachments for small arms fall into two broad categories. Firstly there are those which are integral to the operation of the weapon in its basic role of firing rounds, and either enhance the performance or reduce the signature of the weapon. Examples are muzzle brakes and flash hiders. In the second category are attachments which enable the weapon to fulfil a secondary role and include bayonets and grenade dischargers. It is usually only necessary to attach some device at the muzzle when the designer is trying to meet conflicting requirements. The soldier may well ask for a weapon which is light and short and at the same time produces minimum flash and recoil. These are difficult to reconcile, but with the help of muzzle attachments the designer may get some way towards what is requested.

Muzzle Brakes

A muzzle brake will partially reverse the flow of emergent gases, symmetrically, and in so doing reduce the recoil. The diagram at Fig. 6.1 shows how, in its simplest form, a baffle surrounding the path of the bullet, mounted slightly

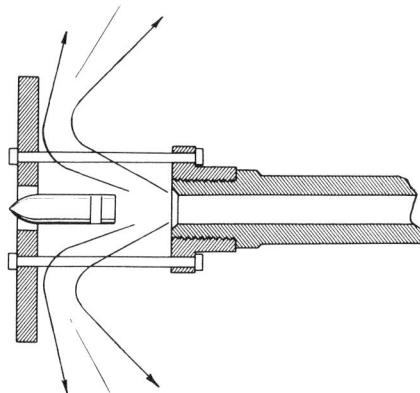

Fig 6.1 A Simple Muzzle Brake

forward of the muzzle, will deflect the gases to the rear. This produces a forward pressure on the weapon directly counteracting the recoil. In Volume II of
this series the subject of muzzle brakes is covered more fully.

The chief problem with a muzzle brake is that it also deflects the blast wave to
the rear and sidewards; to the firer and his comrades close to his flank this can be
an unpleasant and even dangerous increase in the decibel level leading to what is
known in medical circles as acoustic trauma. Consequently a balance has to be
struck between a reduced efficiency in braking force and an acceptable noise level
for the firer. The following table shows the efficiency of various types of muzzle
brakes when fitted to an SLR.

TABLE 6.1 Effect of Muzzle Brake on SLR

Type of brake	Approximate reduction in recoil
90° - one exit	25%
120° - two exits	40%
Flash hider alone	10%

Compensators

It can also be beneficial to deflect the gases asymmetrically. This will result in
an uneven force on the weapon which can be employed to counteract other unwanted
effects. One such effect is the tendency of a burst of rounds to climb up the target, often to the right. By using the gases to deflect the muzzle in the other
direction this component of error in accuracy is compensated for - hence the
name. A typical compensator is that on the assault rifle shown in the photograph
at Fig. 6.2. After the bullet has left the bore the exhaust gases try to expand out
of the muzzle in all directions. Some of them act on the inner surface of the compensator forcing the weapon downwards and to the left.

Fig 6.2 Compensator on Soviet AKM

Recoil Intensification

One of the methods of operation described in Chapter 8 is recoil. Since the available recoil energy from ammunition of 7.62 mm calibres and below is at best only marginally adequate, a recoil intensifier needs to be fitted. The design of recoil intensifiers can vary, but perhaps the best known current example is that in the West German MG3 GPMG, and the diagram at Fig. 6.3 illustrates its method of operation.

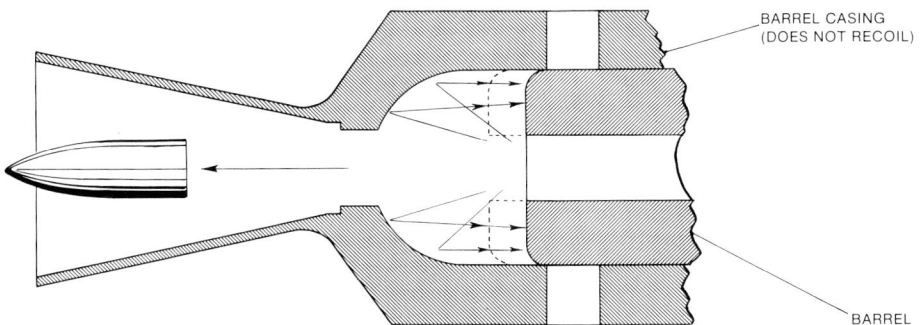

Fig 6.3 Recoil intensifier - 7.62mm German MG3, GPMG

Concealment

The firing position of a small arm can be given away by flash, smoke and noise. The first two are bound up with each other in that the chemical nature of propellants is such that decreasing one giveaway signature will normally cause an increase in the other. The propellant for most weapon systems is therefore designed to minimise the smoke, since the way in which it hangs around makes it a persistent giveaway. The hope is that the much briefer flash will be less revealing, especially if some effort is also made to conceal that.

Flash hiders come in two main types, those with bars and those with cones. The bar type of flash hider consists of three or five equi-spaced metal strips running out from the muzzle. The mode of flash suppression is imperfectly understood, but it probably has to do with the breaking up of the mixed gas and solid cloud that issues from the muzzle, some of which burns in the atmosphere. The cone is more frequently met with nowadays, and features on such weapons as the UK GPMG. The illustrations at Fig. 6.4 overleaf show a circular plate behind the cone of one of the flash hiders. It is designed to reduce the flash for the firer who might otherwise be temporarily blinded at night.

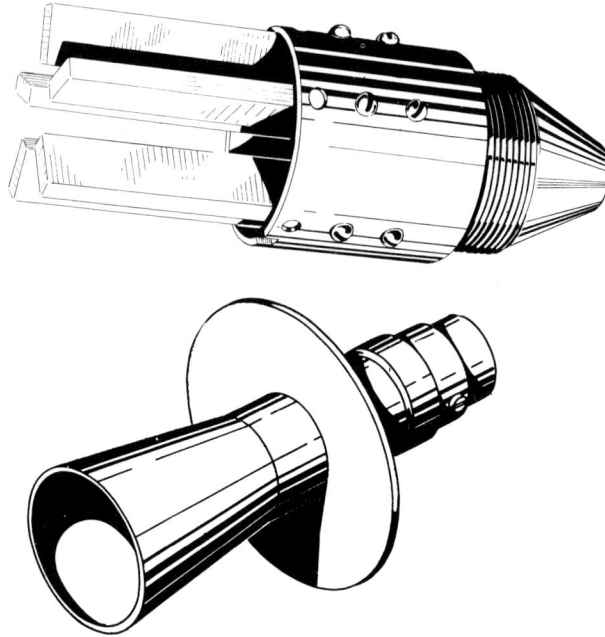

Fig 6.4 Bar and Cone Flash Hiders

The other illustration shown at Fig. 6.5 of a 1925 Browning .303 MG shows a cone which is asymmetrical and therefore serves as a compensator also.

Fig 6.5 1925 model Browning .303 LMG
with combined flash hider/compensator cone

Silencers

Giving away the rough location of a weapon by its noise is generally accepted. More often than not the other noises of battle will drown a small arm report, and on other occasions it is only a well trained and experienced soldier who can use the sound of the report to estimate the location of the firer. Silencers also bring with them disadvantages which make them unacceptable for general use. However, there may be some modern anti-terrorist scenarios which demand such a device. The technical problem is to reduce the velocity of the bullet and the gases to below the speed of sound (340 m/sec). This is not possible with high velocity rounds which often have muzzle velocities over double that speed. However ammunition, such as the 9 x 19 mm round used in many NATO SMGs, is only marginally super-sonic and so presents less of a problem. To reduce the bullet speed the gas pres-sure has to be reduced and this is done by bleeding the gas off through many holes along the barrel length. Reducing the velocity of the exhaust gases is done by a series of baffles extending well forward of the muzzle. The overall result is a much heavier longer weapon as is shown in Fig. 6.6.

The noise of the working parts crashing to and fro has also to be considered. This noise is considerably reduced in a weapon firing with advanced primer ignition as the blow of the working parts against the chamber is cushioned by the early detona-tion of the cartridge. This indicates a further advantage of silencing a weapon such as the UK L2A3 Sterling sub machine gun shown in Fig. 6.6.

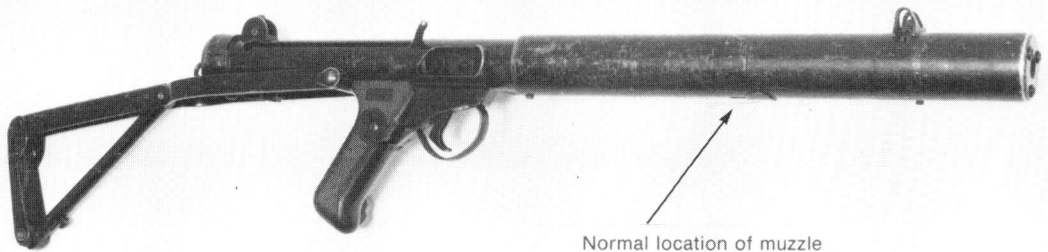

Normal location of muzzle

Fig 6.6 Sterling 9mm SMG with silencer

SECONDARY ROLES

The two principle attachments, which when placed on the muzzle of a weapon give it a secondary role, are the bayonet and grenade dischargers. For the training role there is also the blank firing attachment.

Bayonets

The very closest of close quarter fighting has involved infantry in the use of the
bayonet since the pike went out of service. It came back into prominence in the
trench warfare of World War I, when bullets fired in such cramped conditions
were as dangerous to friend as to foe. The nature of warfare today reduces the
likelihood of bayonet fighting. The most likely scenario is in house to house
clearing, but even then the grenade and SMG are usually preferred. However, an
aura still surrounds the bayonet, and there are emotive arguments that its use in
extremes may be the turning point of the battle. Most infantry requirements for
rifles therefore include the need for fitting a bayonet. This is not always as easy
as it sounds since the modern light and short weapons lend themselves poorly to
fighting with a blade on the front, and other types of attachment may make the
muzzle an inconvenient fixing point. In overcoming this difficulty most requirements accept something more akin to a fighting knife, which can incidentally be
attached to a weapon, rather than the long strong blade of the traditional bayonet.
The modern barrel is frequently too short for the purpose, and there is a real
danger that a heavy blade put to hard use could bend a barrel. To justify such an
attachment at all some armies seek to make the bayonet an all purpose tool, which
can open a can and a bottle, cut wire, perhaps be a weapon adjustment tool, and
may also be used as a fighting knife. Figure 6.7 shows a typical assault rifle
with a fighting knife attached at the front.

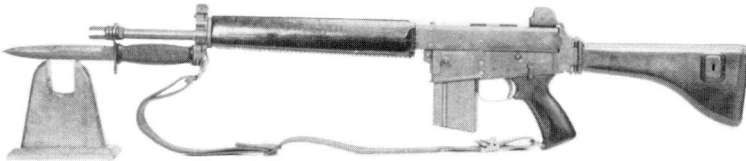

**Fig 6.7 AR 18 rifle with fighting knife
type bayonet**

Dischargers

The value of grenade dischargers is hotly argued. That they confer on an infantryman with no extra weapon, a limited artillery (or at least indirect fire) and
anti-tank capability is undeniable. However, detractors point to the fact that the
grenades have to be carried, their effect is small, the accuracy and range is not
great and at times it is not possible to fire the weapon in its primary role if fitted for grenade launching. Early grenade dischargers around the time of World
War I were cups which took the conventional hand grenade such as the 36 or Mills,
and blew the grenade out with a ballistite cartridge. This meant a separate type
of round, and the normal ball rounds had to be removed before it was safe to fire
the discharger. Furthermore, firing grenades from a cup discharger was

hopelessly inaccurate when it came to try and hit vehicles, especially when moving. One solution to the accuracy problem was to give the grenade tail and fins
which meant that it had to fit over a spigot on the muzzle. The grenade had to
undergo further changes. No longer could an arming handle be allowed to fly off
on discharge. Set-back fuzes to arm the grenade and a graze function to explode
it at whatever angle it hit the target had to be built in. To get accuracy a high
velocity was necessary at discharge, and this, coupled with the weight of the
grenade, led to considerable recoil. Reducing recoil forces to an acceptable level
and at the same time improving accuracy tend to be somewhat incompatible. A
balance has to be struck which results in grenades which do not have a large lethal
area but can be fired out to the required range with sufficient accuracy to be effective.

In some cases the spigot can be integral to the weapon if no other muzzle attachment is to be used, but more often than not it has to be put into place before use,
generally on the bayonet lug. A special sight has also to be brought into play in
some way, since the trajectory of the grenade is so different to that of the bullet.
Such a sight is normally on the back of the spigot. These features can be seen on
the photograph at Fig. 6.8 which shows the UK L1A1 SLR fitted to fire the Energa
anti-tank grenade.

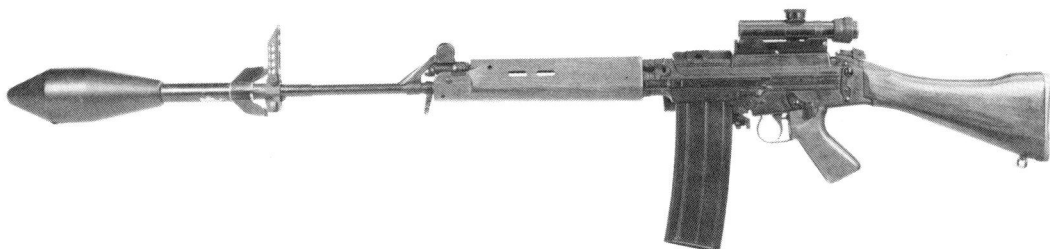

**Fig 6.8 L1A1 rifle with Energa spigot
launcher, grenade and sight**

The need for a special cartridge to propel the grenade can be overcome. Either
a bullet trap is employed or the round passes right through the grenade and in
doing so some momentum is passed to the grenade. A bullet trap can consist of
a series of baffles in the tail of the grenade which successively slow the bullet as
it pierces them, whilst at the same time imparting the bullet's momentum to the
grenade. Before the last baffle is ruptured the bullet and grenade have the same
forward velocity. The Belgian Mecar grenade is an example of a bullet trap
grenade, the bullet being trapped in a metal plug in the spigot. When a bullet is
allowed to pass right through a hollow grenade there may be diaphrams which the
bullet pierces and in so doing imparts energy, or the exhaust gases following the
bullet may be used to propel the grenade.

There is a tendency nowadays among requirements staffs to demand or accept a
separate tube to launch grenades, either as a totally separate weapon, or in
some cases as an add on feature, with a grenade launching tube attached below

the barrel. The American M79 firing the 40 mm grenade is an example of the specialist weapon for this purpose but this is now being replaced by the M203, which is a tube mounted on the M16A1 rifle. A prototype of this system is shown in Fig. 6.9.

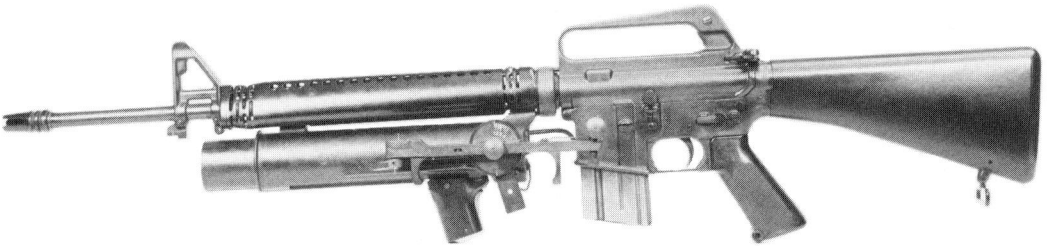

Fig 6.9 M16 A1 rifle mounting M203 grenade launcher

Blank Firing Attachments

In a manually operated weapon a blank round can be fired with no modification to the weapon. Providing sensible safety precautions are observed the materials, gaseous and solid, which emerge from the muzzle are of little danger. As we shall see in Chapters 7 and 8, if a weapon is automatic or semi-automatic a proper firing cycle is needed to make the mechanism repeat its action, and this requires a projectile to go up the barrel, or at least a high gas pressure to be built up. This can only be done in many automatic weapons by using blanks which have wooden or plastic bullets. So that these pose no danger to troops exercising in minor tactics, the bullet has to be broken up before it leaves the muzzle. This is achieved with a Blank Firing Attachment (BFA) which breaks up the wood or plastic into tiny splinters which rapidly lose their energy, and so are not able to cause injury unless fired ridiculously close to another soldier. An alternative blank firing attachment is the type used with the FN FAL 7.62 x 51 mm rifle, which totally blocks off the barrel, and in so doing causes sufficient pressure to build up in the barrel to activate the working parts. This type of attachment is shown opposite in Fig. 6.10.

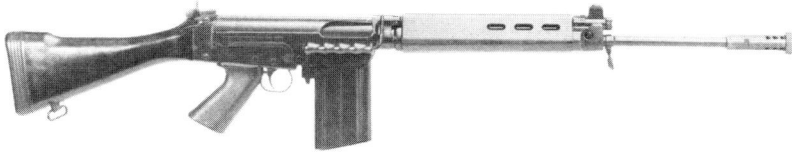

Fig 6.10 FN FAL Rifle with blank firing attachment

Requirements

The requirement for a muzzle attachment, or any additional fitting, to a small
arm has to be most carefully considered. The attachment must be an essential
and not just a desirable feature because fitting it to a weapon may prejudice the
overall performance of the weapon in its primary role. There are disadvantages
in seeking to increase the versatility of a small arm beyond its primary function
of providing the soldier with the means of firing a bullet at an enemy.

SELF TEST QUESTIONS

QUESTION 1   What two broad categories of muzzle attachment are there for small arms?

Answer   ....

....

QUESTION 2   What problem does a muzzle brake bring with it?

Answer   ....

QUESTION 3   Which two problems of weapon signature can muzzle attachments reduce?

Answer   ....

....

QUESTION 4   What will a compensator achieve?

Answer   ....

....

QUESTION 5   What disadvantages attend the use of a grenade discharger?

Answer   ....

....

QUESTION 6   What can be done to improve the accuracy of an anti-tank grenade launcher?

Answer   ....

QUESTION 7   How can the need for special cartridges be dispensed with when firing a grenade discharger?

Answer   ....

....

QUESTION 8   Why is a Blank Firing Attachment needed for an automatic weapon?

Answer   ....

....

QUESTION 9   Describe two varieties of flash hider.

                Answer   ....

                          ....

QUESTION 10   Why are bayonets less likely to feature as a military require-
                ment than in the past?

                Answer   ....

                          ....

ANSWERS ON PAGE 184

7.

Cycle of Operations

For all conventional small arms, and for larger weapons for that matter, there is a sequence of actions which has to take place for each round that is fired. These are:

A round is chambered

The breech is locked (except with most blowback operated weapons)

The firing mechanism is released

The breech is unlocked

The breech block is retracted

The empty case is extracted from the chamber

Energy is stored in a return spring

The empty case is ejected from the gun

The firing mechanism is cocked

A new round is fed into position for chambering

Within this process the cap is struck igniting the propellant, and the bullet is forced up the barrel. With a few exceptions all these actions take place in every weapon, whether manually operated, semi-automatic or fully automatic. How automatic fire is achieved will be covered in a later chapter, but first it is important to review the processes listed above, and these will be described under:

Feed systems

Extraction and ejection

Trigger mechanisms

Locking will be covered in the chapter on Safety.

109

FEED SYSTEMS

The feed system includes:

> The holding device which presents the ammunition to the weapon and trans-
> fers it to the feed mechanism (magazine, belt or strip)
>
> The feed mechanism which takes the round from the holding device and
> positions it for loading
>
> The final process of loading it into the chamber

The first element in these is the magazine, belt or strip which supplies the gun
with the ammunition. The role of the weapon dictates which of these types is
preferred, and it is important to understand the reasons for the choice.

The basic decision is whether portability and rapid reloading is more important
than a sustained rate of fire. In close combat roles such as the assault or winning the fire fight, the infantry must carry its weapons and ammunition and be
able quickly to attach a new supply, preferably with no need to put down the
weapon. This is only feasible with a magazine, which can generally be removed
and stowed with one hand, a new magazine selected and fixed, without letting go
of the weapon with the other hand. Magazines are therefore generally to be found
on hand weapons such as SMGs, assault rifles, rifles and certain LMGs.

If a sustained covering fire is needed, with few breaks for reloading, then a belt
feed is the better choice. Other factors plainly need to be taken into consideration. One of these is serviceability. The section infantryman is likely to subject
his ammunition to much rough handling, not deliberately, but because of the way
he has to move and operate. If he throws himself to the ground and over walls,
crawls through mud and undergrowth, and wades streams, this may all affect the
ammunition, and rounds which are contained in a magazine are better able to
withstand such treatment and still function properly when needed, than those in a
belt.

Modern practice is quite often a mix, with section riflemen carrying magazine
fed small arms, but the supporting gun team having a belt fed weapon. The
practice has been taken a step further with the FN Minimi 5.56 mm LMG, which
is capable, without modification, of accepting either belt or magazine. Certain
other LMGs have this facility, but usually require an alteration to the feed cover,
as does the Russian RP-46. The photographs at Fig. 7.1 opposite show the
Minimi using both a magazine and a belt to supply it with ammunition.

Fig 7.1 Minimi alternative feeds

Magazines

Magazines are described by their shape: box, drum and tubular, with drum being
further subdivided into gun driven (eg the Lewis gun) or spring driven (eg the
Thompson). Tubular magazines are rare since pointed cartridges and centre
caps lead to a danger of one round setting another off during the feeding process.
The common example of a tubular magazine is the shotgun. The photographs
overleaf show both tubular and drum magazine weapons.

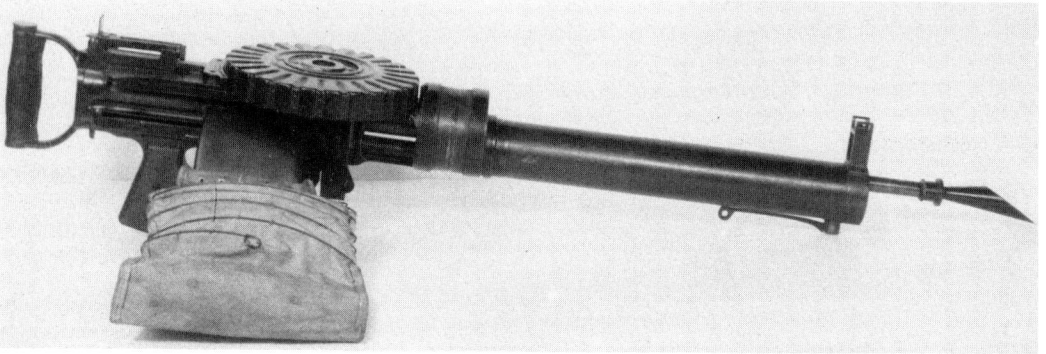

Fig 7.2 Lewis gun with drum magazine.
Note compensator at muzzle

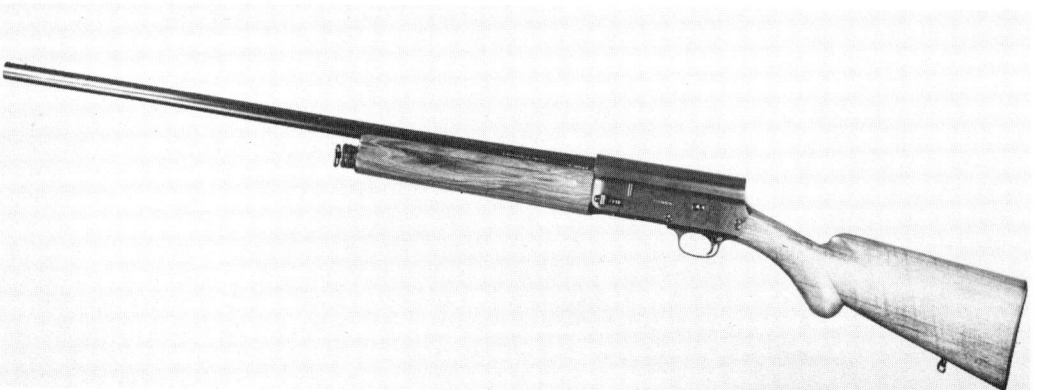

Fig 7.3 Browning automatic shotgun with
tubular magazine

The box magazine may be curved or straight, and have single column loading or staggered, but they all have the basic components: case, platform and spring.

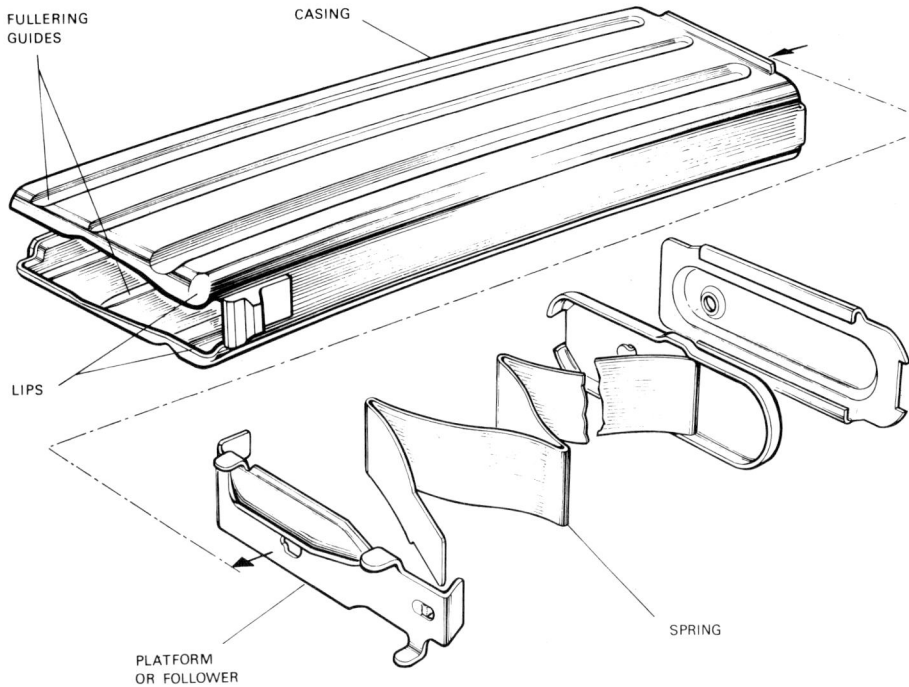

Fig 7.4 Magazine, Box Type - LMG 7.62mm L4 A1

There are several important design considerations affecting the magazine. These include:

Control of the rounds, particularly if in staggered column. The control is achieved by the width of the case, the shape of the platform or follower, and the form of the lips.

Angle of presentation, which is controlled by the lips, and is designed so that the round feeds forward into the chamber without jamming. This also affords a mechanical safety feature, which is covered in Chapter 9.

The spring, which must be strong enough to push the last round up when nearly extended, but not too strong when fully compressed.

The method of attachment. It should be easy for the firer to fit and remove a magazine. The magazine must also be held sufficiently rigidly in position to avoid undue variation in the angle of feed of the rounds.

Position of fitment to the gun. This can be above, below or to the side and each have their pros and cons, which are considered below.

Magazine Position

If fitted above the gun a magazine's spring has some help from gravity, but the magazine stands in the sight line (so the sights have to be offset as shown in Fig. 7.5) and also raises the silhouette, making concealment of the gun harder. A magazine below the gun is out of the sight line, and reduces silhouette, but it is more difficult to change magazines and if it is long enough to take a useful number of rounds, say more than 20, it suffers from interference with the ground. Further, the magazine spring has to overcome gravity. A side fitting magazine upsets a weapon's balance and may make the gun clumsy to handle, but magazine changing is generally quicker and easier. It is the preferred position for many close quarter weapons like SMGs.

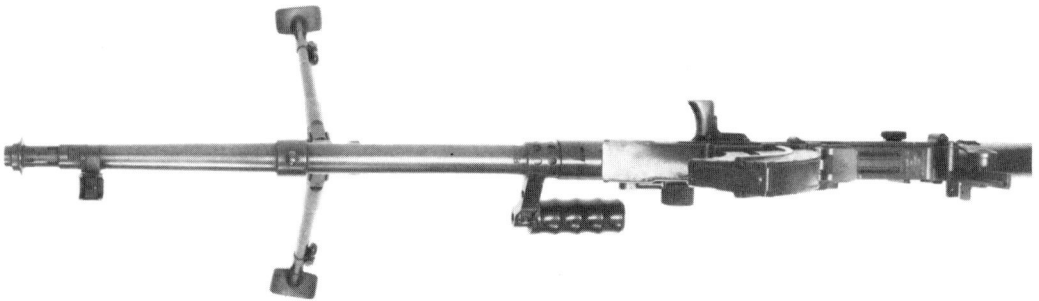

**Fig 7.5 Vickers Berthier .303 LMG with
magazine mounted on top resulting in
offset sights**

Drums

The design and detailed operation of drum magazines will be covered in the handbooks of the obsolescent weapons which use them. In brief, a gun driven drum does not rely on a spring to push the rounds out but requires a complicated gun mechanism of feed pawls. The spring driven drum magazine is similar in operation to the box magazine with a spring pushing a follower behind the rounds. Indeed it was to be found as an alternative to box magazines on some machine guns thus conferring either a sustained fire role or ease of portability to the same weapon.

Magazine Filling

In this era of cheap, non-returnable packaging the thoughts of designers are turning to plastic throwaway magazines as a possible method of ammunition supply. This could be cheap and convenient. Until this happens, however, there will be a need to fill magazines since they are an expensive and integral element of the weapon system, and must be used again and again. Plastic, reusable, see-through magazines are a new development which allow the soldier to see when his magazine needs recharging. Many magazines are filled by hand and a simple drill is used by the soldier to force the rounds against the follower and spring, so that they are correctly staggered and ready for use. Some magazines require chargers or clips. Chargers position the rounds over the magazine ready for the charging process of pushing them in, whilst a clip holds the rounds as they go into the magazine. An alternative to the throwaway magazine is the throwaway clip, actually a factory filled, water-proof, dirt-proof pack. A few magazines require a tool to assist in filling as is the case with the 30 round Sten magazine.

Belt Feed

Belts are more suitable for sustained fire. For an equivalent number of rounds they are lighter than magazines. Their vulnerability to damage, dirt etc means that they are most commonly used in vehicles, for tank coaxial armament for example, but this is not to say that they are not frequently found being used down to infantry section level. Belts come in a variety of forms, namely fabric, fabric with metal strips, metal, metal cum fabric and metal disintegrating link.

Whilst fabric is easier to manufacture it is vulnerable to moisture or rot and inclined to stretch. It provides a cheap, factory filled, expendable supply, which has to be contained in a metal box until required for use. It could not be refilled after use, mainly because of the stretching.

A metal belt is obviously more costly and difficult to make, and is heavier, but it can be relied upon to position the round correctly, can be exposed to the elements and still work well, and can be refilled in the field if need be. Those entirely made of metal are generally counted to be the best belts. Many designers compromise with a fabric metal mix. A lighter cheaper belt results which has an adequate performance and can usually be refilled.

The diagrams overleaf at Fig. 7.6 show a number of different types of belt and indicate the weapon which fired each type.

A FABRIC STRIP BELT—VICKERS M G

B STRIPLESS BELT (ALL FABRIC) VICKERS M G

A,B   .303 in

C,D   7.92 mm

E   7.62 mm

C CONTINUOUS LINK BELT—
   GERMAN M G 42

D METAL CUM-FABRIC BELT—
   BESA M G

E DISINTEGRATING LINK, 7.62mm MK 1—
   CURRENT G PMG

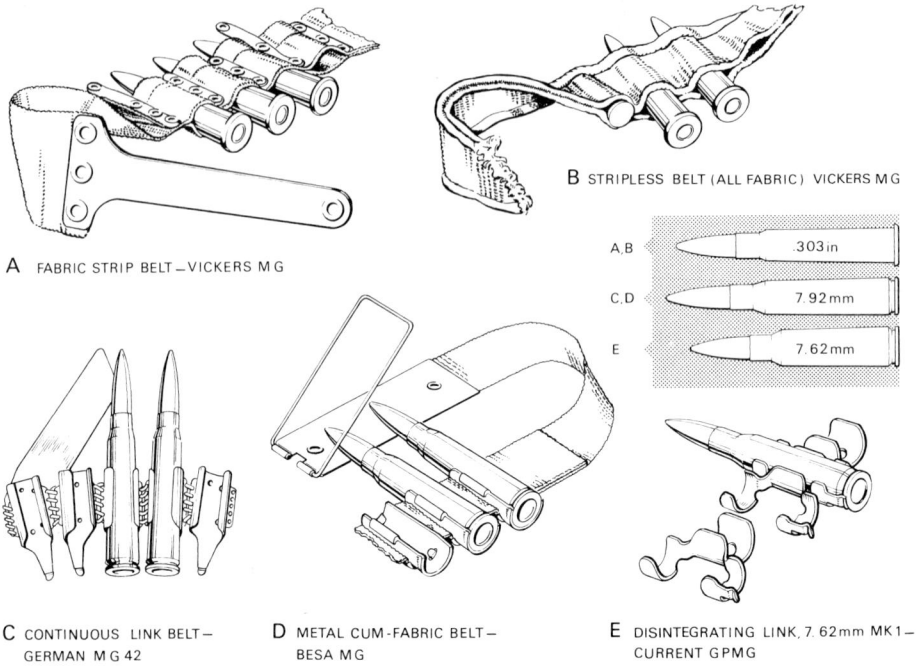

Fig 7.6 Belts and Links

Metal belts are in most common use today, and fall into two main categories:

Continuous metal belt
Disintegrating link belt

Germany developed the continuous metal belt for the MG 42, and the Soviets now use it on the Goryunov MMG M43 and the RPD LMG and PK GPMG.

**Fig 7.7 Soviet RPD with continuous metal belt.
The alternative drum magazine needs
a slight change to the cover**

NATO have opted for disintegrating link belts for most of their GPMGs. It is not intended that disintegrating link be recharged in the field and it is an expensive commodity. It can also lead to problems if the links are not properly collected in an AFV, and could cause stoppages. To save costs the links are recovered as salvage whenever possible on training.

Strip Feed

This form of ammunition presentation is a compromise between a belt and a magazine. The photograph at Fig. 7.8 shows a typical strip feed. Normally the rigid metal strip could only support up to 25 rounds before bending under the weight of the rounds and so leading to stoppages. Strips are now obsolete, but can be seen on outdated weapons in museums.

Fig 7.8 1920 0.303in Hotchkiss with 22 round strip feed

Feed Mechanisms

The magazine presents a round in the feedway ready for the bolt or breech block to feed it forward. Belts (and strips) need a mechanism to pull or push the next round into position for feeding. Such mechanisms may be

    Gun driven

    Spring driven

    or Externally driven

A spring driven mechanism has to have the spring energy stored by the gun's operation, so it is only a variation on gun driven feeds.

Externally driven feed mechanisms are instanced by the hand cranked Gatling gun, or modern externally powered automatic weapons like the Vulcan 20 mm, the 7.62 mm Mini gun and the Hughes Chain gun. Detailed descriptions of the operation of these weapons can be found in the makers' handbooks.

Gun driven feeds are almost universal for such infantry weapons as use belts. Frequent reference is made in other chapters to the energy available in the form primarily of exhaust gases to confer on the weapon an automatic capability. This energy is generally more than enough to operate the actuating levers, slides and pawls which draw the belt through the weapon, and so line up successive rounds ready for feeding into the chamber. Figure 7.9 shows the system on the UK L7A2 GPMG. A very high cyclic rate of fire may not be possible with a simple gun driven feed mechanism. This is one reason for the introduction of externally powered guns.

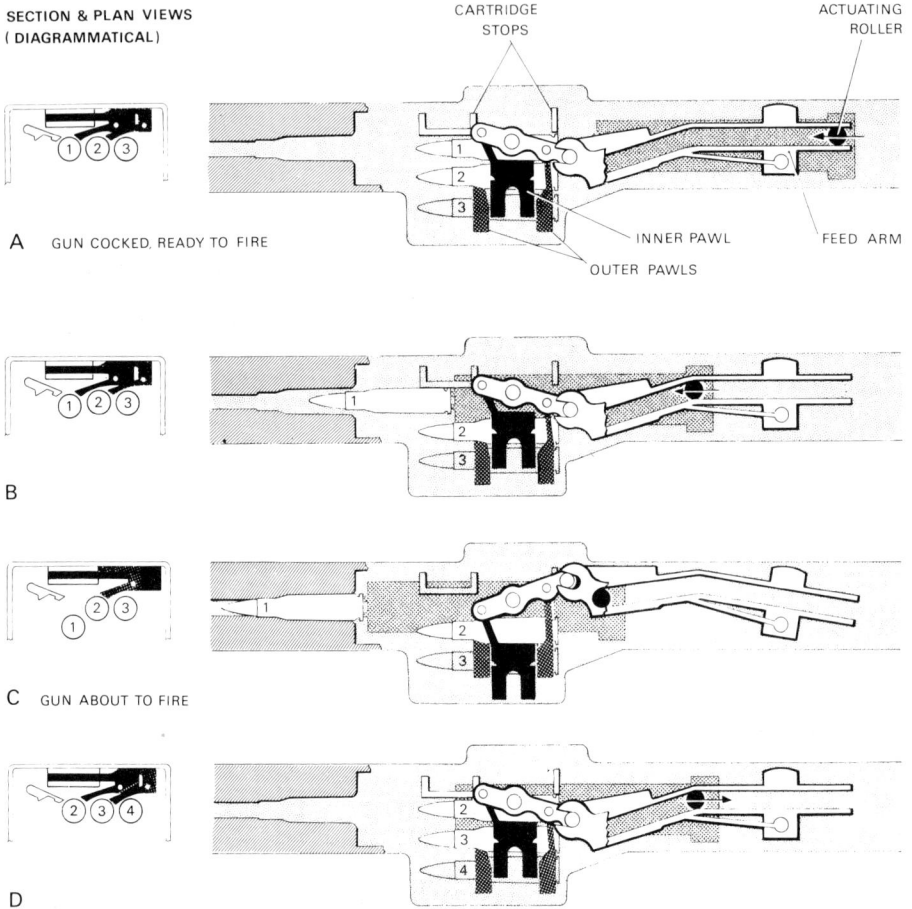

SECTION & PLAN VIEWS (DIAGRAMMATICAL)

CARTRIDGE STOPS

ACTUATING ROLLER

A   GUN COCKED, READY TO FIRE

INNER PAWL   FEED ARM

OUTER PAWLS

B

C   GUN ABOUT TO FIRE

D

Fig 7.9 Gun Driven Feed Mechanism - 7.62mm L7A2 GPMG

Care has to be taken also to ensure that the available power is sufficient to overcome gravity and pull the belt up from the ammunition container to the gun. The possible elevation or depression of the gun on its mounting must also be considered as this may introduce twists in the linkage. Overcoming this extra resistance to motion may overstretch the available power resources. Normally such problems only occur with vehicle mounted machine guns where the ammunition container can be some distance from the gun. It can of course be easily overcome by using a 250 round belt in a box which itself can be clipped to the gun.

Loading the Round into the Chamber

If the magazine or belt feed mechanism has done its job properly the round will be placed ready to be driven into the chamber. As a rule, whilst the breech block was moving to the rear, the round was being moved to the correct position for loading. Now, as the breech block runs forward it rams the base of the bullet, sliding it out of the magazine or belt, and into line with the axis of the bore, so that the nose of the bullet goes into the chamber. Closing the breech completes the loading of the round. In revolving Gatling type guns, a loading lever or ramming piston is needed to perform the same function as the breech.

Holding Open Devices

A holding open device is fitted to most automatic weapons using magazines and has a number of useful purposes. Firstly, the weapon is held open when no more rounds are available and so clearly indicates to the firer that an empty magazine is the cause of the stoppage. By being held open, magazine changing is made quicker and easier as the weapon does not have to be recocked to load a new round. Secondly, in a weapon firing from a closed bolt, a manually operated holding open device is necessary to be able to clear stoppages, to allow the weapon to cool more rapidly and if it has become very hot, to avoid cook off. It is usually the arrival of the magazine follower at the feed mechanism which triggers the holding open device. Belt fed weapons cannot be so easily adapted to a holding open device, however with such weapons there is not the same need for rapid reloading as with the magazine fed weapons which are used in close quarter fighting.

EXTRACTION AND EJECTION

These two expressions are self explanatory, in that they comprise the extraction of the spent cartridge case from the chamber and its ejection away from the gun.

Extraction

Pulling the cartridge case out of the chamber may not be quite so easy as it sounds. Both the chamber and the case expand when the burning of high velocity ammunition propellant creates pressures of about $340MN/m^2$ (20 tons/sq inch). However the yield strength of the chamber steel and cartridge brass differ. The chamber, after expanding, will revert to its previous dimensions, not having exceeded its elastic limit. The cartridge case may more easily pass this limit, and not shrink back to its original size. It is possible for the case to become pressed so hard against the walls of the chamber that extraction is difficult or impossible. This can be prevented by careful control of the upper level of chamber pressure and of the composition of the cartridge brass.

Apart from a few weapons which have parallel sided cases, most cartridges will have some taper. Thus when once dislodged by that first little bit, known as primary extraction, removal is smooth and easy. Primary extraction is usually done by applying some additional leverage at the moment when the breech block starts to the rear, with a device such as a cam shaped lug. A fluted chamber is sometimes incorporated in the design of blowback operated guns. Gas in the fluting is at the same pressure as that in the case, and a smaller area of the casing surface is in frictional contact with the inside of the chamber. These aspects are explained more fully in Chapter 8.

The actual extractor is usually some form of claw which grips the rim at the base of the cartridge. Since considerable force may be needed for primary extraction, and the acceleration forces involved are of a high order, the extractor needs to be robust, but at the same time it has to be kept small. As the claw has to grip round the rim during the feeding stroke, a spring is often used to hold it in position. On the rifle bolt this was achieved in one piece of spring steel attached at two points as shown in Fig. 7.10.

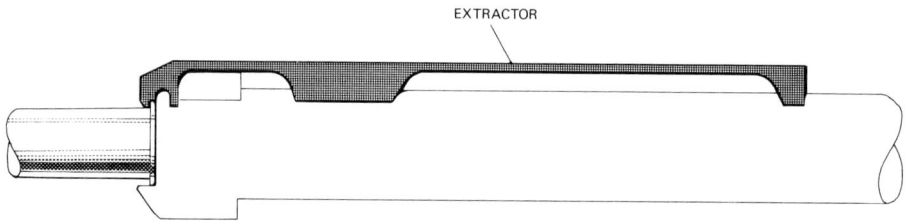

Fig 7.10 One Piece Claw Extractor - No 3 Rifle

More common is the multi piece claw, where spring and extractor are two components.

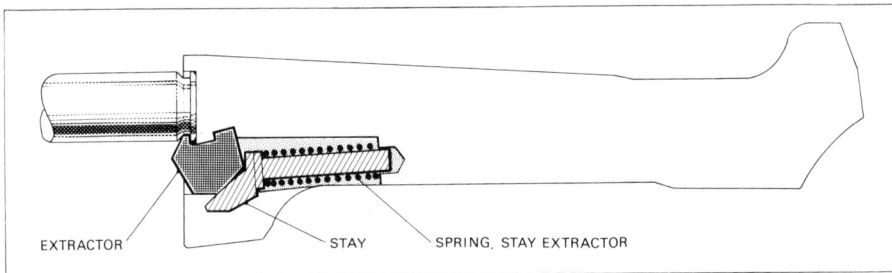

Fig 7.11 Multipiece Claw Extractor - 7.62mm L4 LMG

An alternative, not using a spring, is the T type extractor, on the face of the bolt, where the rim is fed up or down behind the extremes of the T. The Vickers and Browning MGs used such an extractor, and it was particularly necessary as the round had to be pulled rearwards out of a fabric belt before being fed. Weapons which break open like revolvers and shotguns have combined extractors and ejectors. An example is at Fig. 7.12 opposite.

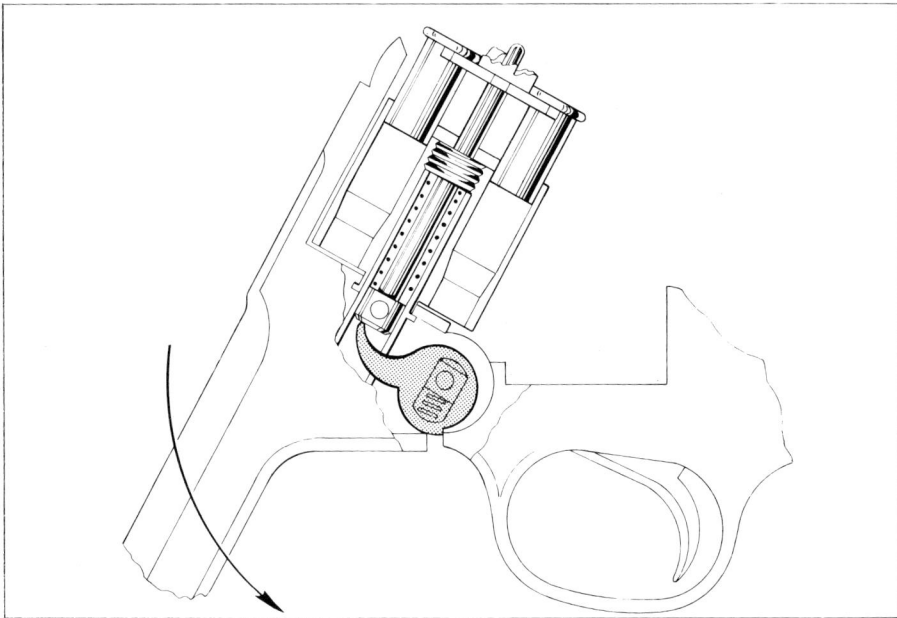

Fig 7.12 Combined Extractor-Ejector - Enfield Pistol

Extractors need to be strong but small and this contradiction in requirement leads to failures. As a result spares of this store are almost invariably carried into action.

Ejection

Apart from the combined extractor-ejector mentioned above, there are three main types of ejector which are illustrated overleaf at Fig. 7.13.

> Fixed
>
> Rocking
>
> Push rod

The fixed ejector lies in wait for the empty case which is being withdrawn by the breech block. At the point where the case is opposite an opening in the body of the gun, its base hits the fixed ejector on the far side from the opening and is forced out through it, pivoting on the extractor. A rocking ejector is pivoted at the centre. As the breech block runs back a projection hits the rear end of the ejector causing the front end to pivot out and strike the side of the empty case. It is a less violent method than fixed ejection, and reduces the chance of distorting the case and so leading to a stoppage.

Fixed

EXTRACTOR   BREECH BLOCK

EJECTOR

Rocking Arm

BOLT HEAD   EJECTOR   PUSH ROD   BUFFER

BOLT BODY

Push Rod

EJECTOR   BOLT

EXTRACTOR

Spring Loaded Rod

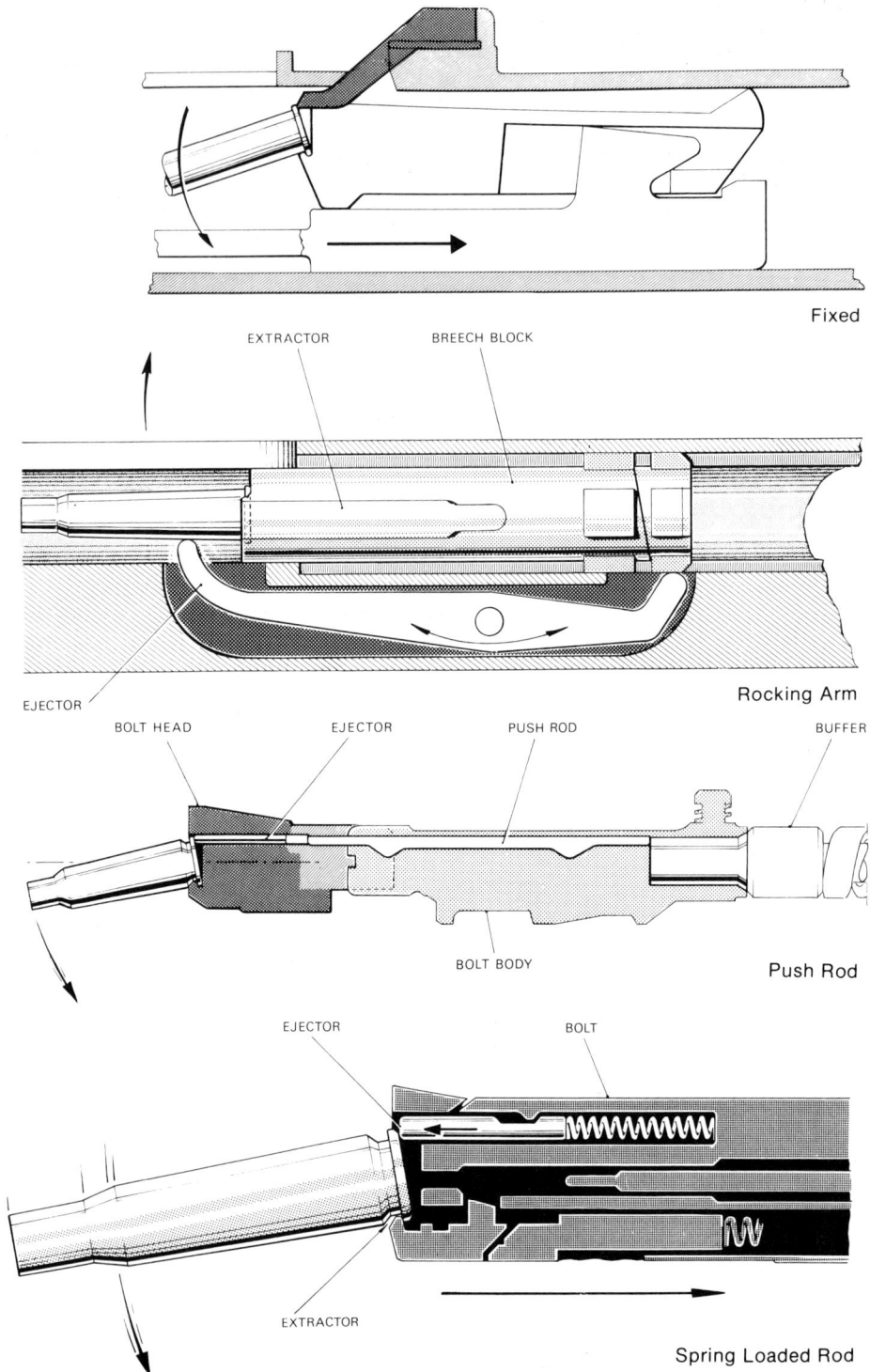

Fig 7.13 Types of Ejector

A push rod ejector acts in some ways like a fixed ejector, namely, through the base of the bolt onto the base of the case. Hitting the side of the base opposite the extractor causes the case to pivot through the ejection opening. The push rod is generally buffered and so is less violent than a fixed ejector.

Two stage ejection is in fact ejection by one of the methods described above, followed by the feeding of the empty case into a tube or container, when operation in a confined space (aircraft or AFV) means that spent cases flying freely about would be potentially troublesome.

Ejectors, though small, are vital components since their failure would result in a weapon stoppage. As they have to sustain severe forces they have to be carefully designed or buffered to cope with the shocks involved.

Reliability

It will be noted that at many times during the cycle the round or case will be moving but is not under positive control. When going forward to be chambered during extraction, and particularly at ejection, the scope for misalignment and a consequent stoppage is considerable. An empty case ejected into the confined space of a vehicle may cause problems outside the weapon, for instance a jammed turret. Reliability is a foremost characteristic demanded of all equipments nowadays and particularly of weapons. This is one reason for the move towards externally powered guns which can confer improved reliability because they control the round through all phases of the cycle.

TRIGGER MECHANISMS

Although the trigger is universally understood for its purpose, and it obviously involves some mechanism, there is in fact an equally vital complementary link in the firing chain which needs to be described and this is the firing mechanism. The firing mechanism applies energy to the cap in the cartridge to initiate it. The trigger mechanism controls the firing mechanism.

Firing Mechanism

Electrical initiation of cartridge caps is quite common in cannons of 20 mm calibre and above, but in true small arms initiation is almost invariably by percussion. The percussive blow is applied to the cartridge cap or primer with sufficient energy to explode it by either a striker or a firing pin. The term striker is usually given to a device which is integrally attached to the bolt, or functions in conjunction with its own spring or the piston post of an LMG. A firing pin will receive its energy from an external source such as a hammer.

An example of a striker is that attached to the bolt of a SMG, whereas a firing pin needing a hammer is to be found on the L1A1 SL rifle. Whether striker or firing pin be used it is important that they provide enough energy to initiate the cap without piercing it and that they stay in contact with the cap during firing so that

cap blow-out cannot occur. Firing pins usually operate through bolts and have to
be made to close tolerances and of high grade material to avoid them bending or
breaking.

Triggers

The link between the firing mechanism and the trigger mechanism is the bent and
the sear. The bent is a recess in the firing mechanism. The sear is part of the
trigger mechanism which engages in the bent. Whilst engaged, firing is prevented. When the trigger mechanism removes the sear from the bent, firing can take
place. Engaging the sear in the bent, usually associated with storage of some
energy in a spring, is known as cocking the weapon.

Operating the trigger is the main method of controlling the firing mechanism. In
the case of some revolvers the trigger will also cock the weapon, but usually the
mechanism needs to be cocked already and the trigger merely releases the
stored energy. Most people are familiar with the conventional trigger as a
hooked lever, but it may comprise a rod linkage, lanyard or push button operating
an electrical circuit, though all these forms are generally to be found on vehicle
mounted weapons.

The design of individual trigger mechanisms is very varied and particularly when
the designer incorporates other facilities. These include safety and selection of
types of fire which are covered in Chapter 9. The principles of a simple trigger
operation are shown below.

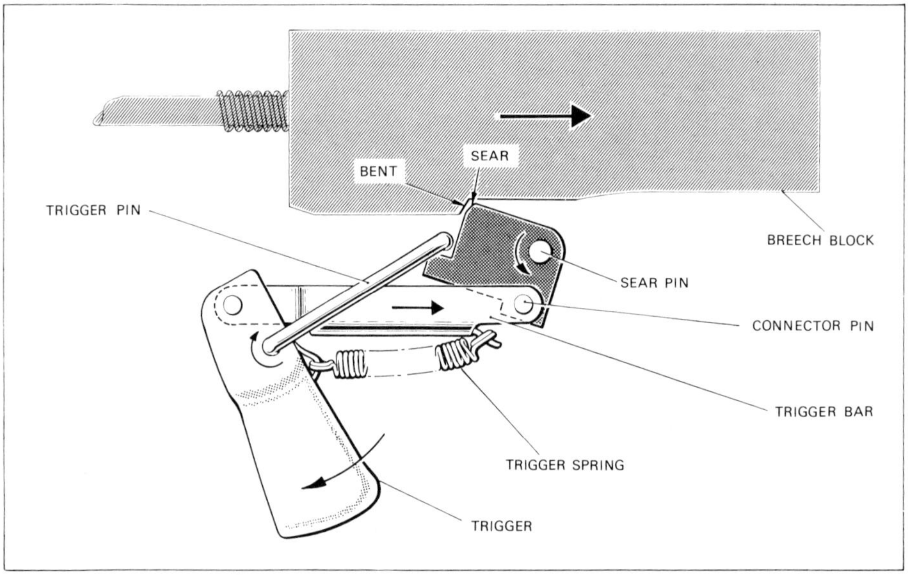

Fig 7.14 Trigger and Firing Mechanism - USA M3 SMG

Open or Closed Bolt

Whether a weapon is of open or closed bolt design is a function of the trigger and firing mechanisms. Advantages accrue to one or the other largely depending on the role of the weapon. With an open bolt design the bolt (or block) is held to the rear during pauses in the firing. The return spring is compressed and the trigger mechanism is holding it in position. When the trigger is operated the bolt flies forward under the influence of the return spring, collects a round, chambers it and supports it for obturation while it is fired. It is a system usually employed with automatic weapons which are required to provide a high volume of fire.

The two major advantages of the open bolt action are firstly that reloading is quicker and easier. The firer has no need to recock, which entails pulling a heavy breech block against a return spring. Secondly, a round is not kept in the chamber for any length of time and cook off is therefore less likely. Since much heat is generated in automatic weapons a round which is left in the chamber for a period after sustained firing may absorb sufficient heat to ignite and cook off. This can be a very dangerous occurrence. Quite apart from inhibiting cook off, an open bolt will allow heat to dissipate more quickly than if the breech is closed.

Weapons firing from an open bolt have a long lock time. The lock time is the interval between operating the trigger and cap strike. This time lag between firing and trigger operation and the small change in the weapon's centre of gravity as the working parts move forward does mean that single shot accuracy cannot be too good. This is because the point of aim may have wandered away from that taken by the firer at the instant of squeezing the trigger.

The closed bolt system gives a much shorter lock time and is therefore most suitable for those small arms from which a high degree of accuracy is important and in which there is no requirement for prolonged automatic fire. As the expression suggests the bolt is closed up to the chamber before firing, and when the trigger is pressed only the striker and spring or the hammer and firing pin have to move.

GETTING OFF THE CYCLE

The series of operations, mechanisms and components which are generally associated with conventional small arms have been described in some detail in this chapter. Designers have been looking for ways to simplify and lighten small arms systems at the same time reducing their cost if possible. In doing so no sacrifice of accuracy, reliability, robustness or other vital characteristics is acceptable.

The best hope of achieving this is to design a weapon which does not have a metal cartridge case. There would be great savings in the weight of the ammunition and ammunition costs would fall considerably without the expense of cases. More rounds per 'magazine' would be a welcome advantage, with anything up to a hundred possibly being mounted along the weapon instead of sticking out from it. Whilst the feeding process is still essential, extraction and ejection no longer are, providing the propellant block is all burnt. The power required to operate the working parts, particularly without primary extraction, would be less. The problem is to produce a block of propellant in which the round is embedded, which can be fed without damage, produces consistent muzzle velocity, and which, without the protection of a case, is not vulnerable to environmental hazards.

Caseless ammunition has been worked on by various countries. In larger calibres, for tank guns with bagged charges, it has been in service for years. In the infantry environment the exposure of the rounds to damage, to damp and to the danger of conflagration is a far more difficult matter to mitigate against. The engineering design of the weapon has been effectively achieved and the West German firm of Heckler and Koch was almost ready with its G 11 weapon to enter it in the NATO small arm trials at the end of the 1970s. There is little doubt that the problems will be satisfactorily resolved and a caseless weapon in the hands of soldiers by the mid 1990s.

POWERED GUNS

Mention has been made of powered guns, which provide a way of getting off the traditional cycle of operation. The reasons for so doing are varied and tend to apply mainly to the larger, cannon calibres, and are enlarged upon in Chapter 10. However, powered weapons in smaller calibres such as 7.62 mm are not only possible, but are in service. The photograph opposite is of the 7.62 mm Hughes Chain gun.

Advantages

The powered gun can offer several advantages over the conventionally operated weapon:

Reliability is increased. The mechanisms which present and feed the ammunition have separate sources of power. These are more than adequate to ensure that belts are lifted from the box or tray where they are stored to the feed opening; they overcome problems of slip, twist or stretch in the belt; the round is fed under control in the proper alignment for entering the chamber; and a faulty round is no longer a cause of a stoppage, since it will be transported through the weapon and ejected as though it were an empty case.

Ejection is controlled. Spent cases are no longer allowed to fly freely from an ejection opening, which can be a hazard in the confines of a vehicle turret or compartment, or an aircraft wing. They are removed under control and can be fed into a tube which then ejects them out of the vehicle.

Clearance drills can be simplified. In the event, hopefully less frequent, of a stoppage occurring, most of the simple causes in a powered gun can be rectified easily.

Inboard space is reduced. The bulky weapon parts required by reciprocating mechanisms, gas ports and pistons, buffers and springs in a conventional weapon may be replaced by a system taking up less space and particularly where this is most important, behind the chamber.

Fig 7.15 The Hughes Chain gun

Barrel changing is quicker. With high rates of fire, changing barrels to alleviate heat problems may be essential. Most powered gun mechanisms can be quickly hinged away from the chamber, allowing the barrel to be withdrawn into the firing compartment.

Toxic fumes are reduced. A weapon relying on gas for its operation usually vents a lot of gas to the rear and consequently into the confined space of a turret if it is so mounted. These gases are toxic and concentrations can only be partially reduced with expensive modifications and attachments. Using external power for the gun's operation means that almost all the gases can be vented outside the firing compartment. Since the ammunition does not operate the mechanism, the gun has a relatively long dwell time, allowing all the gases to go up the barrel.

Increased cyclic rate.   The need for a high cyclic rate, particularly in the anti-
aircraft role, is enlarged upon in Chapter 10.   The rate with a powered gun is
varied simply by the application of more power, and can be raised to a similar
figure to that achieved by a few very specialised designs of conventionally operated guns, offering in excess of 1,000 rounds per minute.   Once power is applied
to the multi barrelled Gatling principle, however, quite startling rates can be
achieved, currently in the region of 6,000 rounds per minute.   The General
Electric 7.62 mm Minigun which is shown below in Fig. 7.16 is capable of producing such a volume of fire.

**Fig 7.16 General Electric powered guns
of the Gatling type**

Disadvantages

The main potential disadvantage to a powered gun is that, deprived of power, it
will not work. Thus a vehicle which is a mobility casualty may also be, at least
in part, a fighting casualty because the gun cannot fire. Nor can the gun be used
in a dismounted role. Some coaxial or pintle mounted machine guns can, if they
are conventionally operated, be dismounted and taken into action. This may not
be possible with a powered gun. Some have a facility for hand cranking, so that
they go on firing without electrical power, though inevitably at a slow rate.
Battery packs can be used as the power sources for these guns in the event of
vehicle power failure, to avoid continuous engine running, or possibly, to facilitate
dismounting.

Application of Power

The extent of the application of power varies. As indicated previously, it can
consist merely of assistance in presenting ammunition to the feed system, it can
then extend to actual feeding under control, and culminate in a process of round
transportation throughout the equivalent of the cycle of operation. One design
which uses external power for all these aspects is the Hughes Chain Gun designed
by the Hughes Company of California.

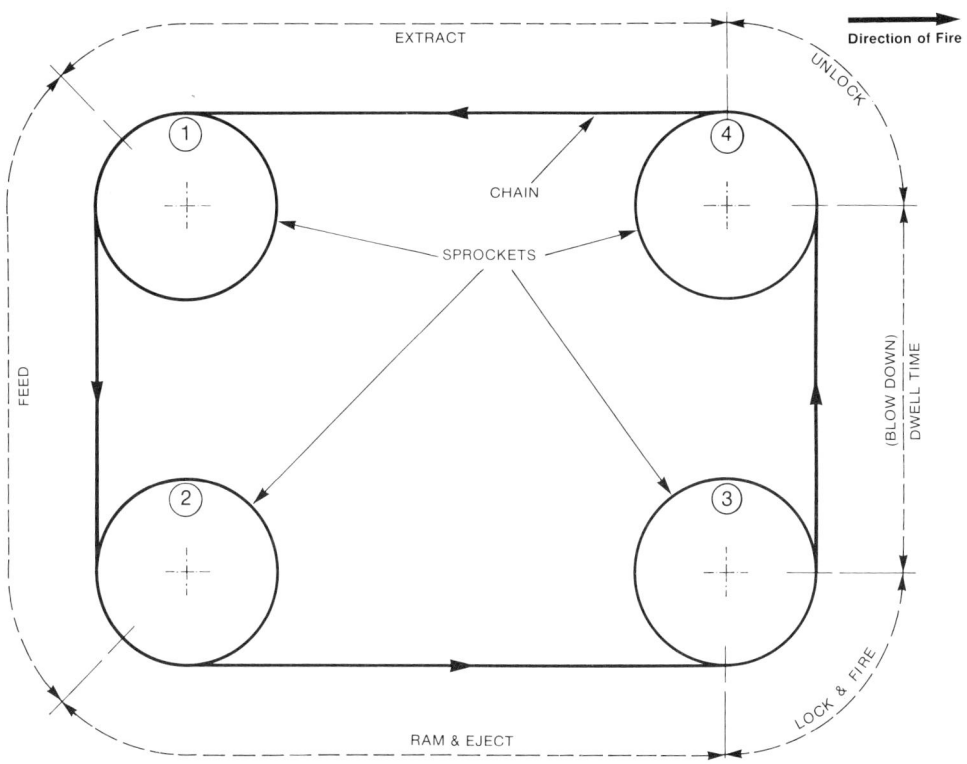

Fig 7.17 Chain Gun Cycle

The Chain Gun is clearly so called because of the mechanism which is driven by a continuous link chain like that on a motor cycle. The diagram on the previous page at Fig. 7.17 gives an outline of the way in which the circulating chain takes the mechanism through the various processes of the cycle. This feature of taking the round through the system with 'positive control' illustrates the intent to make full use of the potential advantages accruing to this type of weapon. Sophistication in weapon systems suggests that we shall see many other developments with similar objectives in mind, particularly if another manufacturer's claim, that these weapons will be cheaper to produce than conventional weapons, proves to be true.

SELF TEST QUESTIONS

QUESTION 1   What main factors govern the choice between magazine and belt
              for an automatic weapon?

              Answer   ....

                       ....

QUESTION 2   What are the arguments for and against a magazine being
              mounted below a gun?

              Answer   ....

                       ....

QUESTION 3   What are the three main types of magazine?

              Answer   ....

                       ....

                       ....

QUESTION 4   What are the two main categories of metal belt?

              Answer   ....

                       ....

QUESTION 5   Why is a holding open device useful on a magazine fed weapon?

              Answer   ....

                       ....

QUESTION 6   What types of feed mechanism are there?

              Answer   ....

                       ....

QUESTION 7   What are the three types of ejector?

              Answer   ....

                       ....

                       ....

QUESTION 8   Where do you find the bent and the sear, and how do they
                interact?

                Answer   ....

                            ....

                            ....

QUESTION 9   Why is open bolt operation favoured for automatic weapons?

                Answer   ....

                            ....

QUESTION 10   What parts of the firing cycle will caseless ammunition avoid?

                Answer   ....

                            ....

ANSWERS ON PAGE 184

8.

Methods of Obtaining Automatic Fire

INTRODUCTION

The initial energy needed for the automatic operation of a weapon comes either from the ammunition or from some outside source. The earliest automatic weapons, such as the Gatling and Gardiner machine guns, required the firer to turn a handle connected to the working parts which then went through the cycle of operations described in Chapter 7. Modern externally powered guns, such as the Hughes Chain Gun are driven by electric motors. To avoid unwieldy power sources, hand held weapons make use of energy from the propellant which would otherwise be wasted.

Propellant energy can be used in three different ways. Firstly, in a process called blowback, the pressure in the bore acts directly on the empty cartridge case forcing the breech block backwards. This compresses springs which are used to push the working parts forward again. Secondly, when the design of the weapon is such that the breech block is not free to move at the instant of firing, high pressure gases act indirectly on the working parts, often via a piston. This is termed gas operation. Thirdly, the weapon can be designed to make use of the effects of recoil.

The efficiency of all three methods, and of blowback in particular, is dependent on the movement of the empty cartridge case while there is still some gas pressure in the bore. However for much of the time when pressure is high the case cannot move or should not be allowed to move out of the chamber. Therefore it is important to study what happens to the cartridge case after the cap is struck.

The Cartridge Case after Cap Strike

In the first fraction of a millisecond after the propellant starts to burn, the pressure forces the cartridge case to expand radially at the neck which is its weakest point. The bullet is then free to move down the bore. The gas cannot escape back past the case because of the close fit between the neck of the case and the chamber

walls.   This process is called rear obturation.   During this very short period the
case can move back slightly to fill what is termed the cartridge head space.   The
diagrams at Fig. 8.1 show the cartridge head space of a rimmed and a rimless
round.   One important point to note is that in order to avoid excess cartridge
movement the breech block should be firmly in contact with the base of the round.

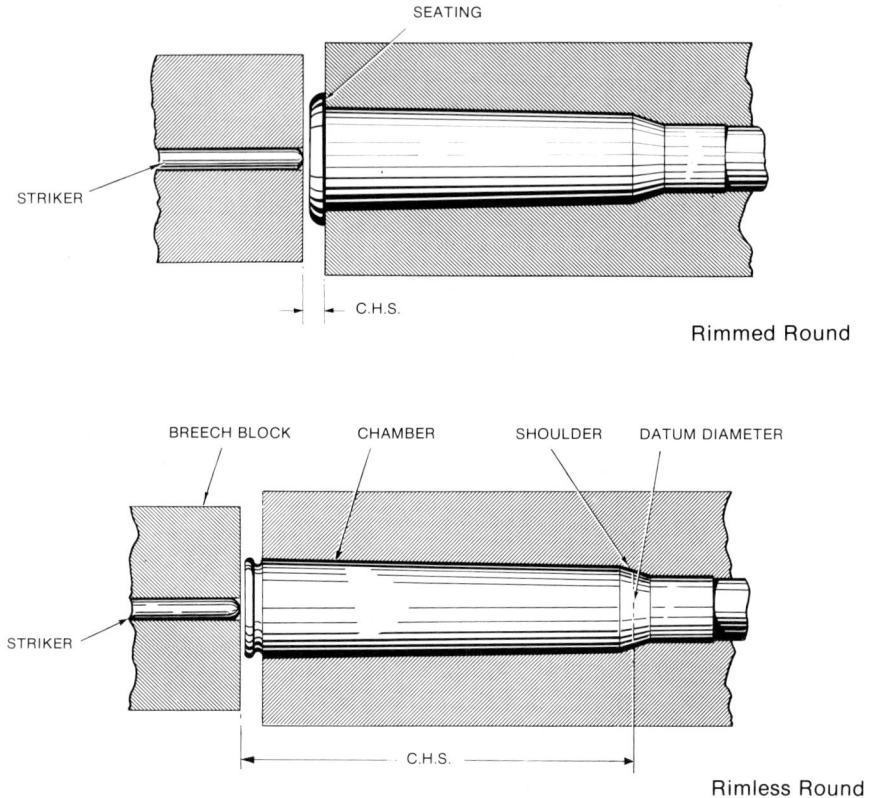

SEATING

STRIKER

C.H.S.

Rimmed Round

BREECH BLOCK   CHAMBER   SHOULDER   DATUM DIAMETER

STRIKER

C.H.S.

Rimless Round

Fig 8.1 Cartridge Head Space

During the next few milliseconds while the bullet moves out of the barrel, the gas
pressure is high.   This high pressure pushes the case outwards against the
chamber walls and backwards against the face of the breech block.   When there is
a close fit between the case and the surrounding metal then the cartridge is not
damaged even when, as in the case of all high velocity ammunition, the pressure
exceeds the tensile strength of the case material.   If there is too much cartridge
head space two types of damage may occur.   These are shown diagrammatically
in Fig. 8.2.

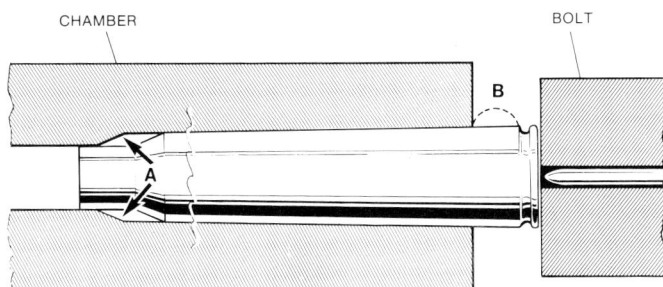

Fig 8.2 Results of excessive cartridge head space

If the space between the neck and the chamber at A is too great, the case may shear. If too much of the thin case wall protrudes at B, the case may bulge or even burst.

Under the pressures produced by high velocity rounds the cartridge case material is capable of greater expansion than the chamber walls. As a result the case is virtually welded to the chamber walls and is incapable of any movement. If the breech face does not provide a rigid support for the base of the round the case is stretched longitudinally and shears.

When the pressure drops the case contracts and is then free to move backwards. If it does so while the pressure still exceeds the tensile strength of the case material the unsupported side walls will bulge. The thickness of the cartridge wall is therefore important. For NATO 7.62 x 51 mm ammunition a safe length of protusion of an unsupported case is approximately 3 mm. For 20 mm ammunition this length can be increased to 6 mm as the case walls are much thicker. These measurements are of critical importance to the design of blowback operated weapons.

BLOWBACK

In the case of the blowback operated weapon the energy for the cycle of operations comes from the effect of gas pressure on the empty cartridge case. The weapon is designed with a breech block which is neither locked to the barrel nor to the body of the weapon so that it can be pushed back by the empty cartridge case.

The diagram at Fig. 8.3 shows that for a necked cartridge case the effective backwards force is less than for a parallel sided case. The latter are therefore more suitable for use in blowback systems.

PRESSURE PRODUCING
TENSION IN CASE

BOLT

PRESSURE PRODUCING MOTION

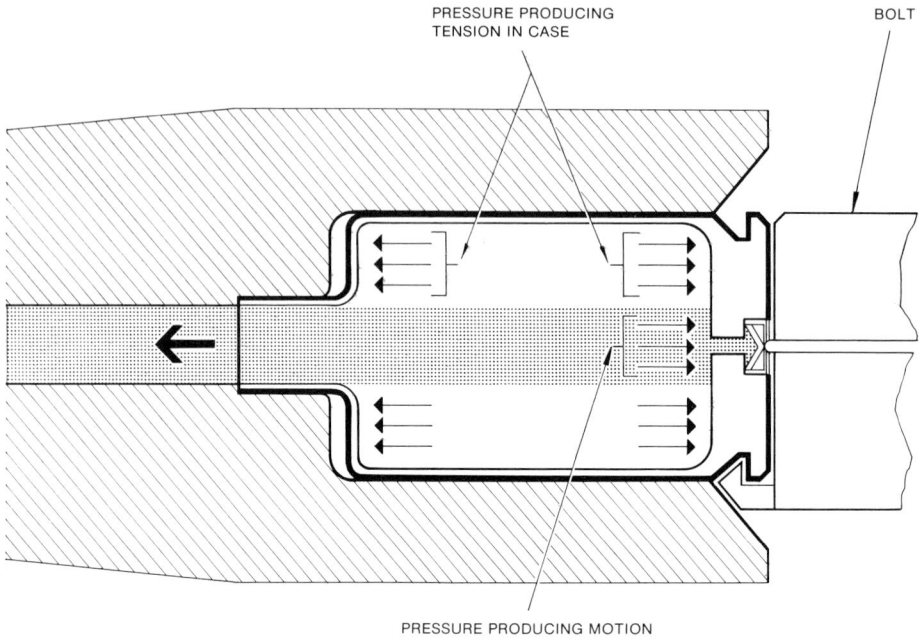

Fig 8.3 Pressure inside necked cartridge case

Blowback can only be effective in the short period after the case is able to move
back with safety and while there is still sufficient pressure in the bore to force
the breech block back against the spring.   Rear obturation is still necessary during this time and is best provided by a parallel sided case.   A problem encountered with the use of the necked case is the fact that the weakest part of the case,
the neck, is the slowest part to contract when the pressure falls.   Therefore
some of the limited available pressure is required to overcome additional friction
forces.   This is a further reason in favour of the use of parallel sided cartridge
cases in blowback weapons.   Unfortunately parallel sides limit the amount of propellant that can be fitted behind a bullet in a round of convenient length.   High
velocity ammunition therefore invariably uses necked cartridge cases.

High velocity ammunition is held fast to the chamber walls by the high pressure
produced so that no rearward movement is possible.   In many blowback designs
it is essential that some movement takes place at high pressure.   The reason for
this will be shown later.   Therefore blowback operated weapons either use low
powered ammunition which does not hold the case too firmly against the chamber
walls or, as described later, have some form of lubrication between the case
and the chamber.   Lubrication also helps overcome the additional friction which
occurs when necked cases are used.

Both high and low powered ammunition can be fired from blowback operated
weapons.   The propellant can be contained in either parallel sided or necked

cartridge cases. As a result there are three different categories of blowback systems called simple blowback, advanced primer ignition and delayed blowback.

Simple Blowback

In simple blowback the bolt (often referred to as a breech block) is stationary at the instant of firing. The gas pressure forces the bullet and the bolt to move simultaneously. The inertia of the bolt, coming mainly from its mass, the stiffness of the return springs and friction, are all that prevents the backwards movement of the cartridge from occuring too soon for safety. Because of this, working parts have to be excessively heavy unless very low powered ammunition is used. Even so the practical difficulties of producing a safe simple blowback system force the designer to reduce barrel length in order to minimise the period of high pressure. As a result the range and accuracy of the low velocity bullets fired from a weapon using this system are generally inadequate for the battlefield.

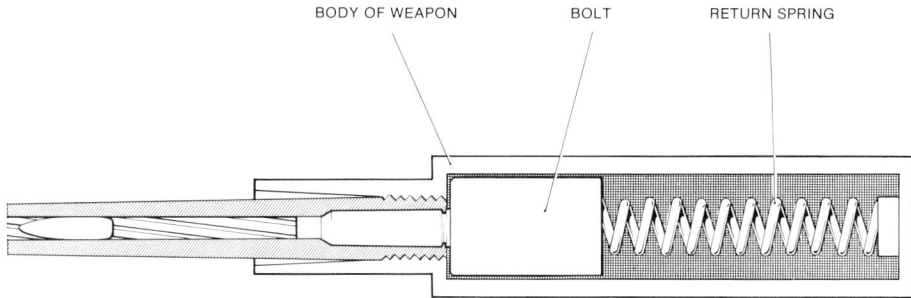

BODY OF WEAPON   BOLT   RETURN SPRING

Fig 8.4 A Typical Simple Blowback System

The photograph at Fig. 8.5 shows the Czech Skorpion which operates on the simple blowback principle. It fires a low powered 0.32 in (7.65 mm) round out to an effective range of approximately 100m against an unprotected man. The weapon is relatively inaccurate at all but point blank range when firing bursts.

Fig 8.5 Czech 7.65mm Skorpion Machine Pistol

Advanced Primer Ignition

Instead of cap strike occuring when the breech block is stationary, it is possible
to design weapons which fire while the breech block is still moving forward.   This
is known as advanced primer ignition.   In addition to forcing the bullet down the
bore, the gas pressure slows down, stops and reverses the movement of the
breech block and return springs.   A weapon using this system can fire more
powerful ammunition without being excessively heavy.

The UK L2A3 Sterling sub machine gun, which weighs 3 kg, operates on the advanced primer ignition principle.   It fires the 9 x 19 mm low velocity bullet which
is effective against unprotected targets at 300m.   The parallel sided case is
crushed up when it enters the slightly smaller chamber which helps to prevent
overstretching the case.   The cap is struck by a fixed firing pin fitted to the
breech block.   To ensure that cap strike takes place while the breech block is
still moving forward, the chamber is designed to provide sufficient resistance to
the forward movement of the case so that the firing pin strikes the cap while the
case is still entering the chamber.   The diagram at Fig. 8.6 shows the position
of the breech when the bore pressure is at its maximum.

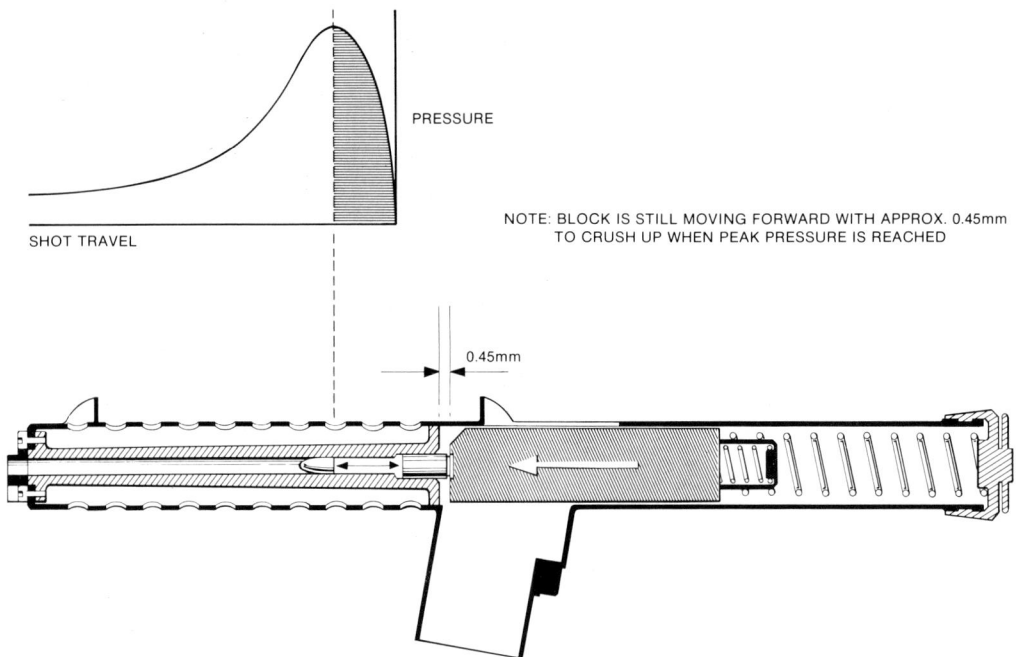

PRESSURE

SHOT TRAVEL

NOTE: BLOCK IS STILL MOVING FORWARD WITH APPROX. 0.45mm
TO CRUSH UP WHEN PEAK PRESSURE IS REACHED

0.45mm

Fig 8.6 UK L2A3 Sterling - Advanced Primer Ignition

The moment of cap strike is critical for advanced primer ignition systems.   If it
occurs too early the case may be burst; if it occurs too late both the weapon and
the case may be damaged.   This is less of a problem when low powered ammunition is used but special design features have to be included in weapons firing high

velocity rounds. Some of these precautions are best illustrated by the UK World War II Naval 20 mm Polsten cannon which is shown in Fig. 8.7.

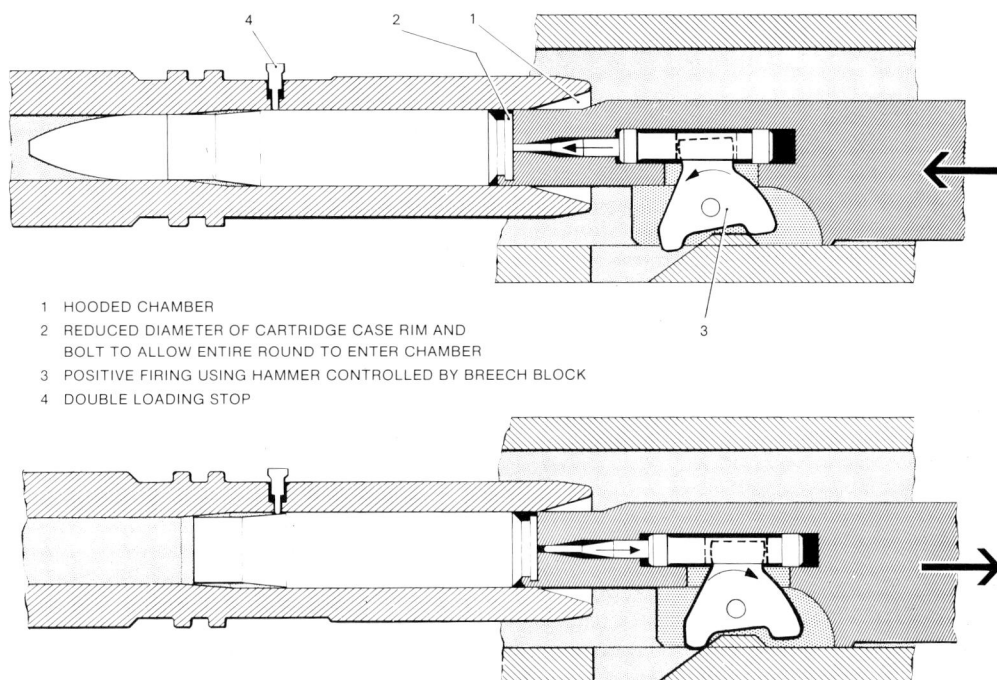

1 HOODED CHAMBER
2 REDUCED DIAMETER OF CARTRIDGE CASE RIM AND
   BOLT TO ALLOW ENTIRE ROUND TO ENTER CHAMBER
3 POSITIVE FIRING USING HAMMER CONTROLLED BY BREECH BLOCK
4 DOUBLE LOADING STOP

Fig 8.7 Polsten 20mm Cannon Safety Features

Firstly, the ammunition was lubricated with grease so that the round could move freely in the chamber even when the pressure was high. Secondly, the base of the cartridge case and the front of the breech block were made small enough to fit inside the hooded chamber. The round was thus fully supported for far longer than in a system using a conventional chamber and the additional time for safe movement of the case helped to reduce the weight of the breech block to about 18 kg. It would have weighed over 200 kg if the weapon had operated on the simple blowback system. Thirdly, the round could only be fired when the hammer tripped a cam. This could only occur when the breech block was moving forwards and when the case was fully supported in the chamber. Fourthly, there was a double loading stop to prevent a fused round from being fed into the chamber in the instance of any obstruction.

Weapons operating on the advanced primer ignition principle are unlikely to be capable of providing accurate single shot fire as they must fire from the open bolt position. The movement of the relatively heavy breech block is likely to disturb the firer's aim.

Delayed Blowback

Delayed blowback allows for accurate single shot fire because the weapon using this system fires from a closed bolt. It is only possible to use a closed bolt in conjunction with the blowback principle if the backward movement of the breech block is delayed until the pressure is low. A simple blowback weapon which fired the NATO 7.62 x 51 mm round would require a bolt of about 20 kg to provide sufficient resistance. When a delay mechanism is included in the design the mass of the bolt can be reduced to less than 1 kg. Figure 8.8 shows the current West German rifle, the G 3, which operates on the delayed blowback principle.

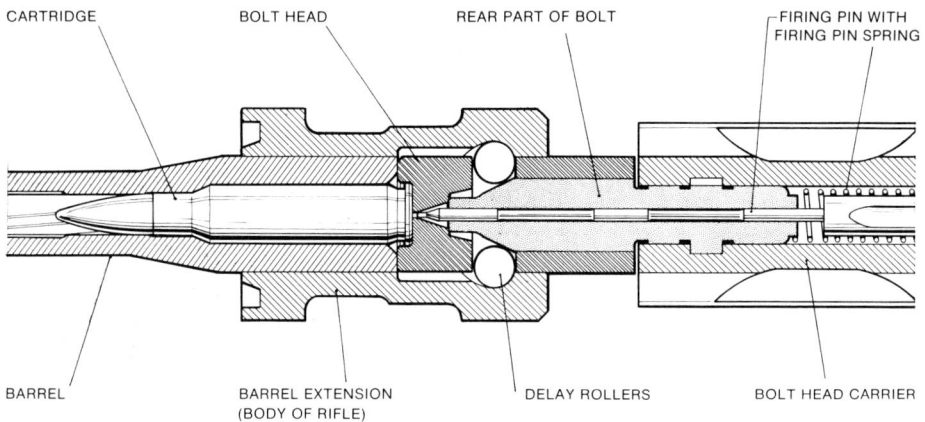

CARTRIDGE   BOLT HEAD   REAR PART OF BOLT   FIRING PIN WITH
                                                                FIRING PIN SPRING

BARREL   BARREL EXTENSION   DELAY ROLLERS   BOLT HEAD CARRIER
                (BODY OF RIFLE)

Rifle loaded, ready to fire

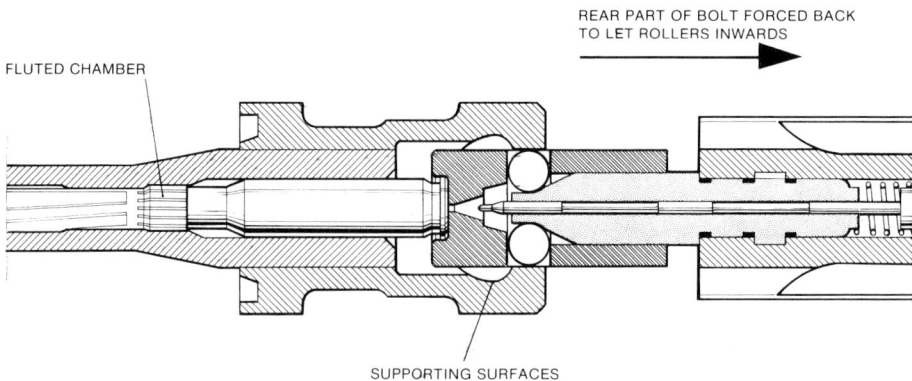

REAR PART OF BOLT FORCED BACK
TO LET ROLLERS INWARDS

FLUTED CHAMBER

SUPPORTING SURFACES

Rifle fired, rollers fully in

Fig 8.8 West German G3 Rifle - Delayed Blowback

By the time the gas pressure has driven the bolt backwards and so forced the rollers into the main part of the bolt it is low enough for the empty case to be extracted safely. The resistance to rearward movement has by then decreased so that the residual bore pressure can supply enough energy to the bolt to operate the rifle. The required delay is achieved by the strength of the springs and the careful design of the shape of the rollers and the recesses in the bolt and body of the rifle.

The G 3 rifle fires the high velocity 7. 62 x 51 mm round which has a necked case. The case must be lubricated otherwise the extra friction makes extraction very difficult if not impossible. To avoid the abrasive problems of oil and battlefield dirt, the case is lubricated by the propellant gases. This requires the chamber to have grooves or flutes cut in the forward part. Equal gas pressure on either side of the case reduces the friction forces when the case is withdrawn. Rear obturation occurs further towards the base of the case than it does in normal chambers. The diagram at Fig. 8.9 shows the important features of a fluted chamber. In the G 3 rifle there are flutes in the chamber extending 40 mm back from the front.

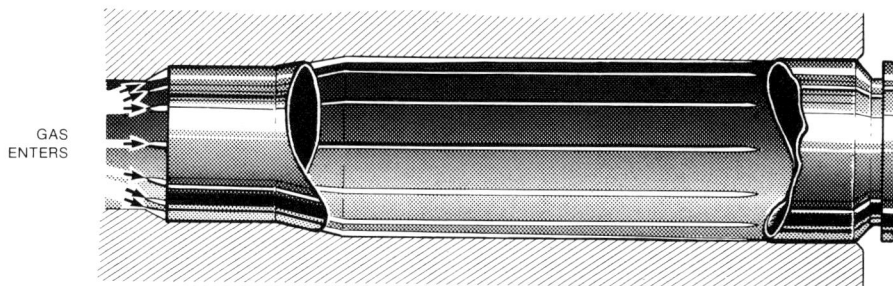

GAS ENTERS

LONGITUDINAL SECTION ALONG CENTRE LINE

Fig 8.9 Fluted Chamber

Blowback Summary

Although blowback operated weapons are relatively simple to manufacture, great care has to be taken with their design to ensure that the empty cartridge case only moves when the bore pressure is at the right level. The margin for error is small so that dirt in the working parts has a marked effect on the weapon's performance. Under battlefield conditions weapons become rapidly affected by dust, grit and fouling so these weapons are designed to have sufficient power to operate when dirty. The ejection of the empty cartridge case from a clean weapon can be fierce. The power which operates blowback weapons cannot be regulated to meet varying conditions. Many of these problems can be overcome by using recoil forces to supply the power to operate the weapon.

RECOIL OPERATION

A recoil operated weapon has its bolt and barrel locked together at the moment of cap strike. They then both move backwards relative to the rest of the bodywork of the gun.   The distance over which they move when locked together gives rise to the two different classes of recoil systems.   Long recoil defines the situation when the recoil distance is greater than the length of an unfired round and short recoil when the distance is less than an unfired round.

Long Recoil

A schematic representation of a long recoil system is shown in Fig. 8.10.

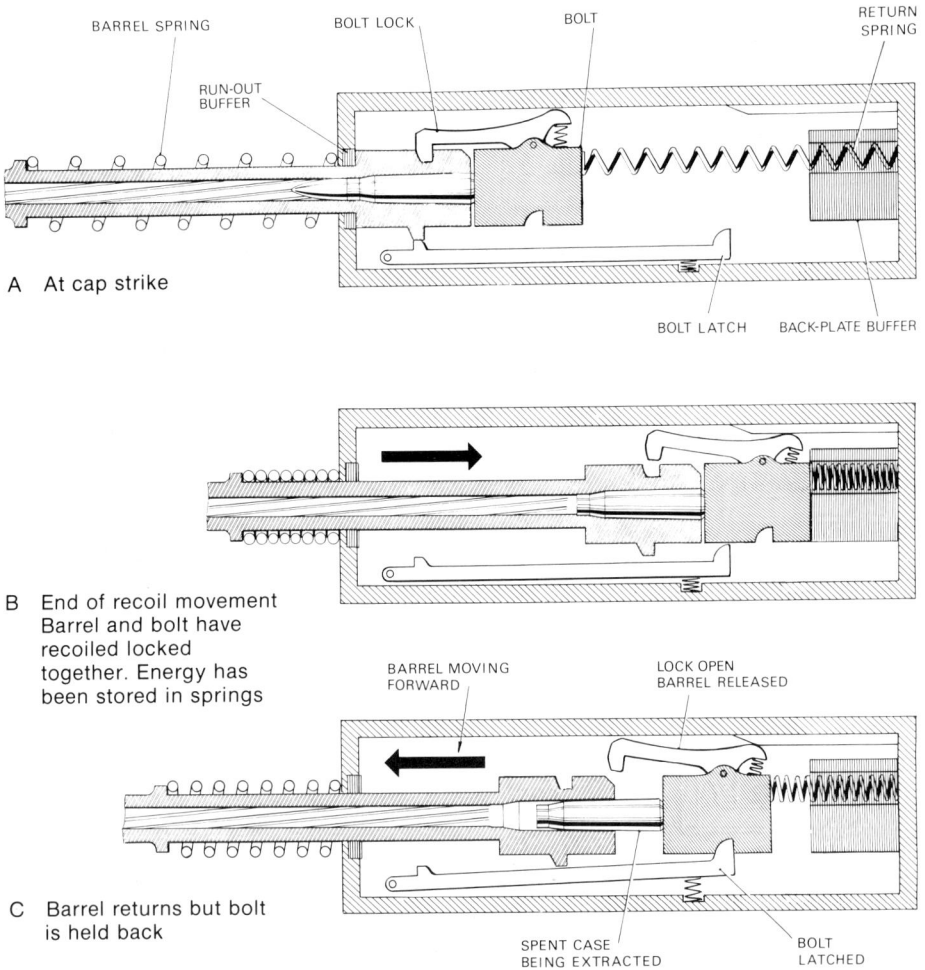

A   At cap strike

B   End of recoil movement Barrel and bolt have recoiled locked together. Energy has been stored in springs

C   Barrel returns but bolt is held back

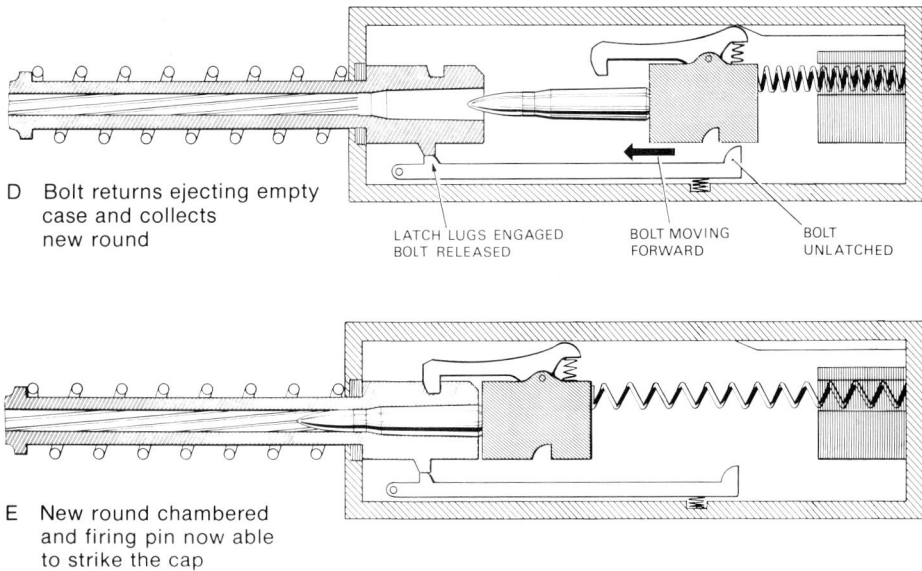

D Bolt returns ejecting empty
   case and collects
   new round

LATCH LUGS ENGAGED   BOLT MOVING   BOLT
BOLT RELEASED   FORWARD   UNLATCHED

E New round chambered
   and firing pin now able
   to strike the cap

Fig 8.10 Diagrammatic layout of a Long Recoil System

While recoil is occuring the pressure in the bore has fallen to a safe level so
there is no likelihood of the empty cartridge case being damaged when it is withdrawn from the chamber. The return springs are storing energy but no other
part of the cycle of operations has taken place. Long recoil weapons must therefore have a slow rate of fire. The force of the recoil is spread over a long time
so that the impulse on the mounting, called trunnion pull, is low. This makes it
possible for long recoil weapons such as the UK 30 mm Rarden cannon to be fitted
on light mountings on lightly armoured vehicles. This system has some disadvantages. Firstly, the balance of the weapon and firer is disturbed by the large
change in the centre of gravity as the bolt and barrel move back between rounds.
Secondly, the rate of fire has to be slow enough to allow barrel vibrations to damp
out. Thirdly, the barrel has to have accurate bearings. All these factors in-
hibit the use of long recoil weapons as rifles or light machine guns. Most of these
problems can be minimised by using the short recoil principle.

Short Recoil

The barrel and bolt of short recoil operated weapons only move back a very short
distance before they separate. In the case of the Browning machine gun this
distance is approximately 6 mm. A short recoil weapon can have a high rate of
fire and the movement of the barrel is unlikely to have much effect on its
accuracy. Therefore it can be used as a light or medium machine gun.

An inherent problem with all recoil operated weapons is that less than one per-
cent of the propellant energy is turned into recoil. Therefore there is some
difficulty in obtaining sufficient power from ammunition of calibres of 7.62 mm
and below. These low calibre weapons need devices to give them additional

power to provide a high rate of fire, to raise long belts of ammunition and to operate when dirty. The extra power can be provided in three ways. Firstly, gas can be deflected back at the muzzle to increase the force on the barrel and bolt. This device is known as a muzzle or recoil intensifier. The one used on the West German MG 3 GPMG is shown in Fig. 8.11.

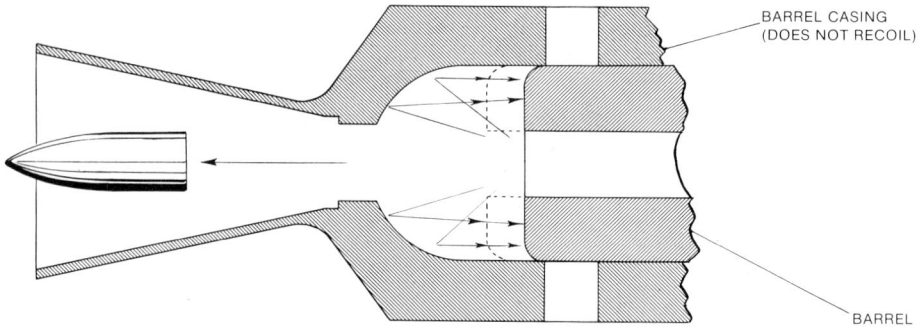

Fig 8.11 Recoil Intensifier - 7.62 German MG3, GPMG

Secondly, the bolt can be unlocked from the barrel when there is some residual pressure in the bore. Blowback can then be used to push the bolt backwards. In the MG 42 (Spandau), which is the forerunner to the current West German GPMG the MG 3, the cyclic rate of fire could be significantly increased when the strength of the springs in the locking system was adjusted so that unlocking occurs rapidly after firing. Thirdly, some of the momentum of the barrel can be transferred by a mechanical device to the bolt immediately after the two have separated. This device is known as an accelerator. Its efficiency depends upon its shape and the ratio of barrel to bolt weights. In practice it is possible to design an accelerator which increases the bolt velocity by up to 50%.

The diagram at Fig. 8.12 shows the schematic layout of a short recoil weapon.

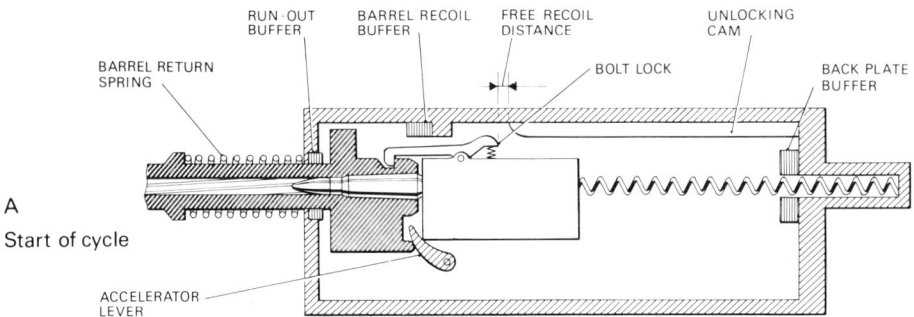

B

Bolt unlocked;
acceleration starts

C

Acceleration
completed

D

Barrel rebounding
from recoil buffer;
bolt rebounding from
back plate buffer

E

Bolt loading
fresh cartridge

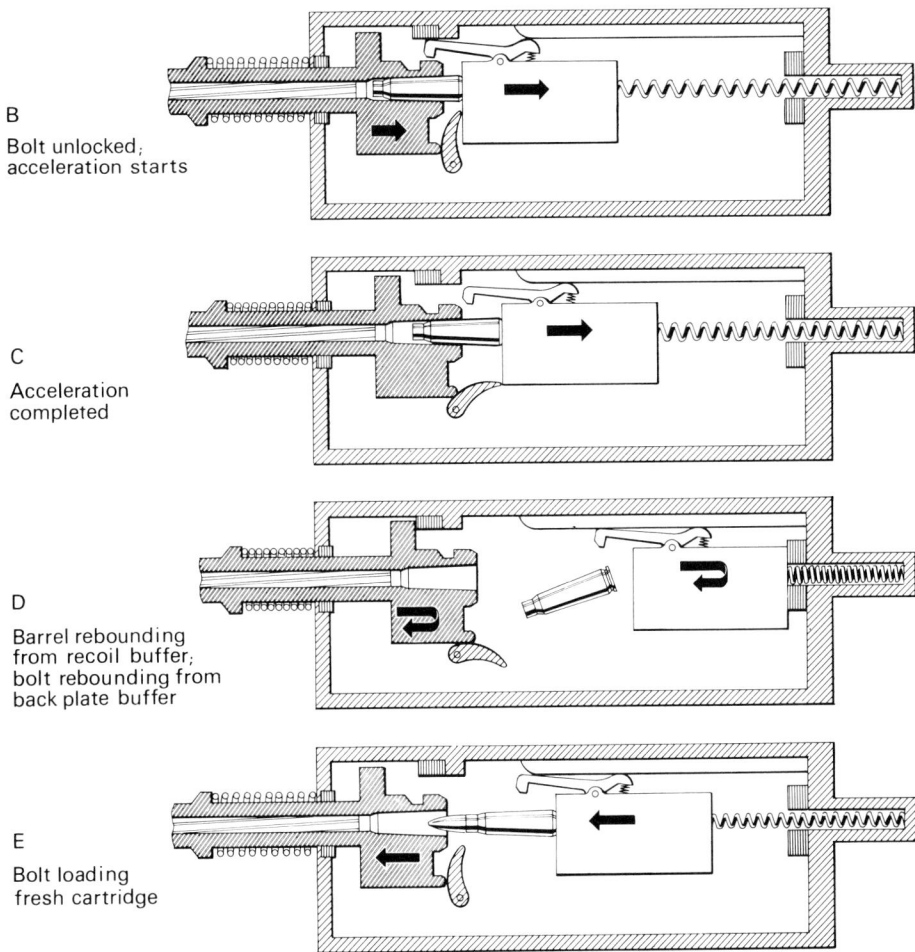

Fig 8.12 Diagrammatic layout of a Short Recoil System

GAS OPERATION

The high pressure gas in the bore is the source of energy for the third method of
obtaining automatic fire. Unlike the blowback system, gas operated weapons
have bolts which are locked to the body of the weapon. The bore pressure cannot
therefore directly force the bolt backwards. Instead the gas is tapped off from
the barrel through a gas port to operate a mechanism which not only unlocks the
bolt but also thrusts it backwards. As there is so much energy in the propellant
gases, gas operated weapons have an ample reserve of power to meet most
battlefield requirements. Figure 8.13 shows the main components of a typical
gas operated weapon.

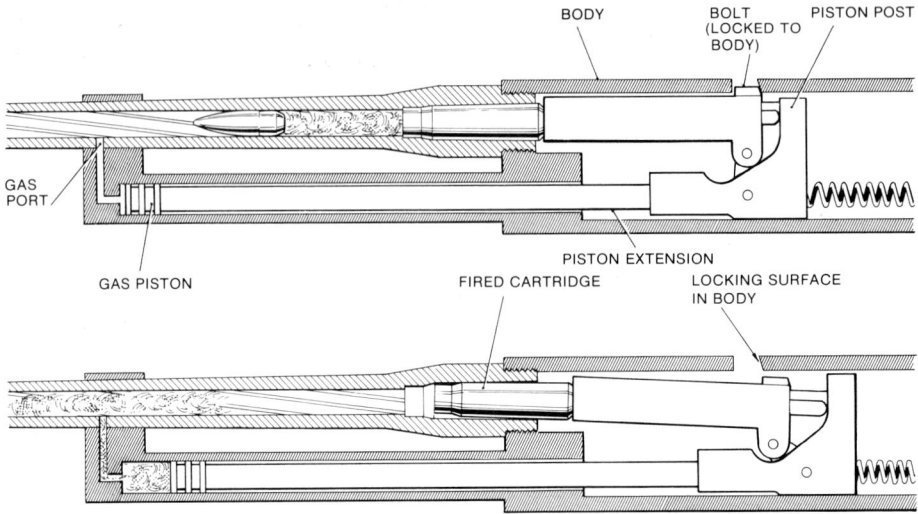

Fig 8.13 Typical Gas Operated Gun

The distance of the gas port from the bolt has a significant effect on the design
and performance of the weapon. If the gas port is close to the breech the bolt is
likely to be unlocked very quickly after the bullet is fired, so producing a high
rate of fire. However, too rapid unlocking may result in the extraction of the
empty case taking place at high pressure with a consequent safety hazard. If the
gas port is near the muzzle there is less erosion but carbon fouling becomes a
problem. In addition to the fouling problem there needs to be a long linkage to
the bolt which may prove awkward to engineer. In practice gas ports normally lie
between the two extremes. Their exact location depends on the power needed to
provide an acceptable rate of fire without incurring excessive erosion or fouling.

Once it has passed through the gas port, the high pressure gas can either operate
a piston, as shown in Fig. 8.13 or exert sufficient pressure in an expansion
chamber behind the bolt to operate the weapon. The piston can either operate
through a long stroke, always remaining attached to the bolt, or it can give the
bolt a short impulsive blow. The diagram at Fig. 8.13 shows a long stroke
piston. Figure 8.14 shows the short stroke system used in the American M1
Carbine.

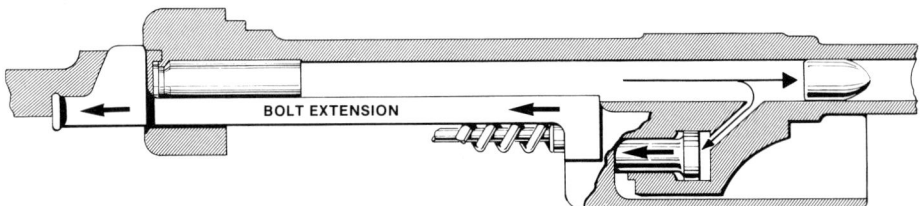

Fig 8.14 Short Stroke Piston - American M1 Carbine

There is no piston in the American Armalite rifle. The gas is fed into a chamber between the bolt and the bolt carrier. It forces the bolt carrier backwards, unlocks the bolt and thrusts it backwards. The diagram at Fig. 8.15 shows the main features of the system.

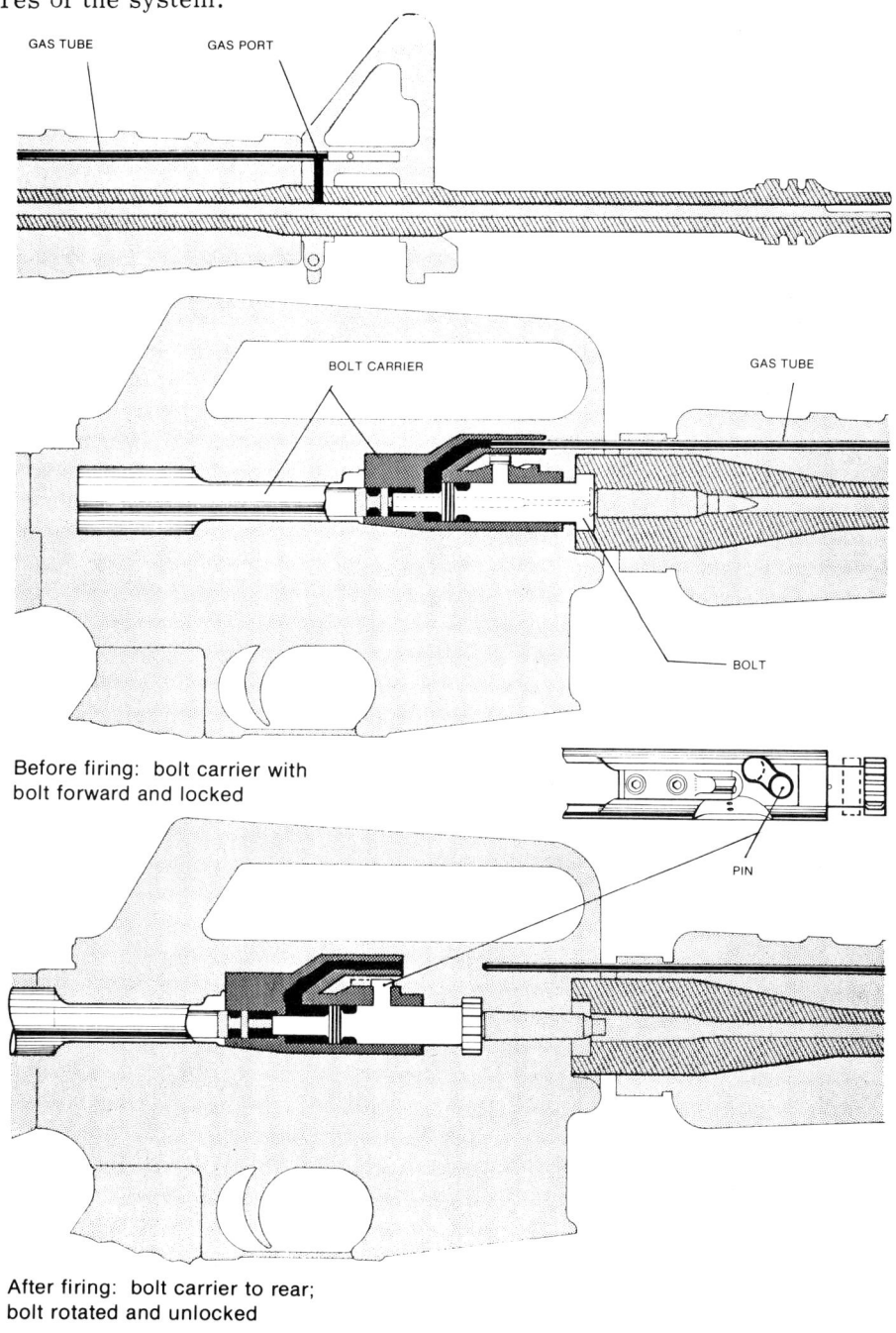

Before firing: bolt carrier with
bolt forward and locked

After firing: bolt carrier to rear;
bolt rotated and unlocked

Fig 8.15 Direct Gas Action - Rifle M16

The amount of gas needed to operate a weapon varies according to the battlefield conditions. However the gas port should only be adjusted in order to maintain the correct cyclic rate of fire. If this rate is significantly varied then the weapon is likely to stop firing because cases may not be ejected cleanly. The dimensions of the ejection opening are based on the movement of the working parts taking place at the correct cyclic rate of fire. Control of the amount of gas is achieved in the following three ways. Firstly, the diameter of the gas track can be varied as shown in Fig. 8.16.

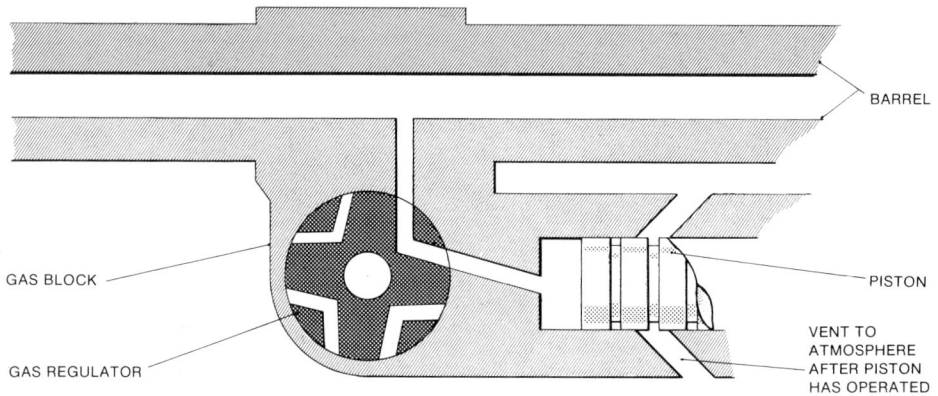

Fig 8.16 Variable Gas Track - Bren, LMG

Secondly, it is possible to vary the size of the hole which allows the gas to escape to the atmosphere as is shown schematically in Fig. 8.17. By reducing the size of the hole more of the gas which has passed through the port is used to power the piston and vice versa.

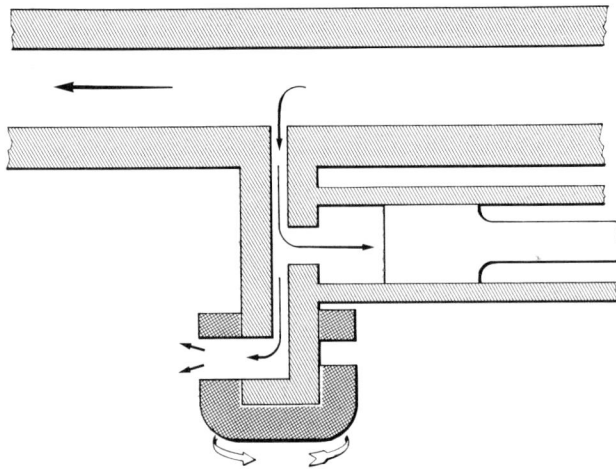

Fig 8.17 Exhaust to Atmosphere System

Thirdly, there is a self compensating system which is used on the American M60 GPMG. Figure 8.18 shows that the gas enters a hollow piston and expands into the gas cylinder. In doing so it forces the hollow piston against the operating rod which is attached to the bolt. As it moves backwards the supply of gas to the hollow piston is cut off. The amount of gas which enters depends on the resistance to motion of the piston and operating rod. Theoretically, as resistance increases more gas is able to enter the piston so that there is more power available to operate the weapon.

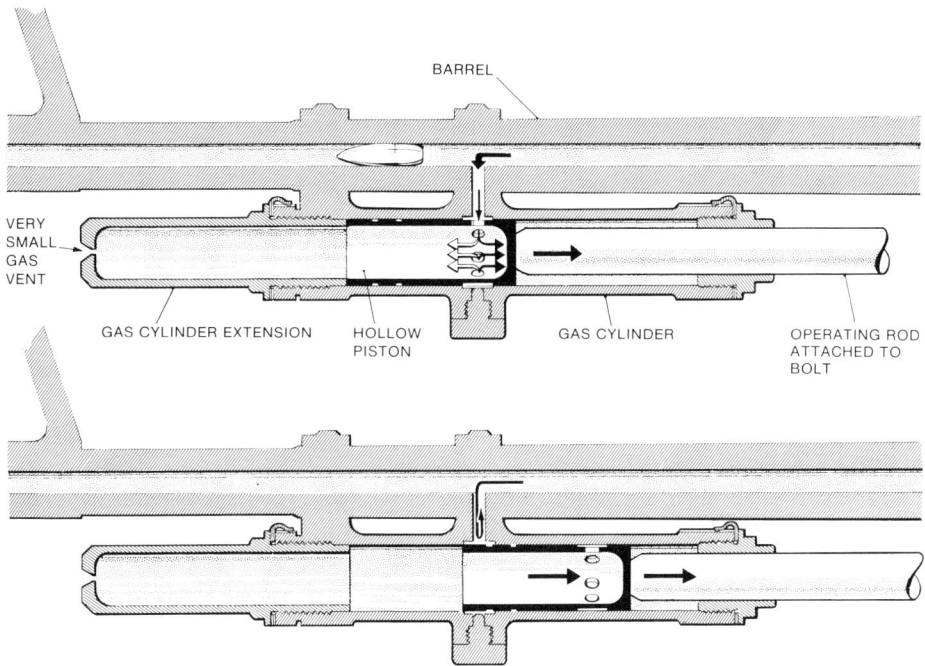

Fig 8.18 Constant Volume Regulator - USA M60 GPMG

COMPARISON OF THE THREE SYSTEMS

At Table 8.1 there is a comparison of the main features of the three ammunition dependent systems used to obtain the automatic operation of a weapon.

TABLE 8.1 Comparison of the Main Features of Gas, Recoil and Blowback Systems

	GAS	RECOIL	BLOWBACK	
			API	DELAYED
Advantages	1. Lightness of working parts 2. Ample controllable power	1. Reliability (no fouling) 2. Few toxic fumes to rear at low cyclic rates of fire	1. Simplicity 2. Cheap 3. Robust	Not quite so good as an API for simplicity, cheapness or robustness
Disadvantages	1. Fouling 2. Erosion 3. Toxic fumes	1. Heavy moving parts 2. Lacks power in small calibres (7.62 mm) 3. Inflexible power supply	1. Heavy moving parts 2. Inaccuracy in hand held weapons 3. Short effective range 4. Fouling 5. Toxic fumes	1. Inflexible power supply 2. Fouling in chamber and body 3. Toxic fumes
Applications	SLR LMG GPMG	Pistol GPMG	SMG	SLR LMG GPMG HMG

HYBRID SYSTEMS

Not all weapons have an easily categorised system of operation. For example some weapons use blowback forces to provide the energy needed to complete the cycle of operations, yet they fire from a locked breech. The breech is unlocked by some device which uses either recoil or the high pressure bore gas to release the locking mechanism. The energy provided by the unlocking mechanism is insufficient for completing the cycle of operations which is then powered by the residual bore pressure; in other words the main energy source is blowback action. The Hispano Suiza Type 820L 20 mm cannon is an example of such a weapon, as is the spotting rifle on a UK light anti-armour weapon. Both use gas pressure to unlock the breech; the US World War II Johnson .30 in LMG uses recoil forces to complete the unlocking process.

SELF TEST QUESTIONS

QUESTION 1   What is the meaning of the term Cartridge Head Space and what is
the effect of too much Cartridge Head Space?

Answer   ....

....

QUESTION 2   What is the main difference between a blowback operated weapon
and one which is either recoil or gas operated?

Answer   ....

....

QUESTION 3   What feature can be found in the chamber of all modern blowback
operated weapons which fire high velocity ammunition?

Answer   ....

....

QUESTION 4   Why are hand held weapons which operate on the advanced primer
ignition principle inherently inaccurate?

Answer   ....

....

QUESTION 5   Why is a short recoil operated weapon likely to have a higher
cyclic rate of fire than a long recoil operated weapon?

Answer   ....

....

QUESTION 6   Why do many recoil operated weapons have recoil intensifiers?

Answer   ....

....

QUESTION 7   What factors must be considered when deciding the location of
the gas port on the barrel?

Answer   ....

....

QUESTION 8   What is the effect of increasing the size of the exhaust vent in the
              gas regulator of the UK L7A1 GPMG?

              Answer   ....

                        ....

QUESTION 9   What is the most suitable ammunition dependent method of
              operation for a VMMG?

              Answer   ....

                        ....

QUESTION 10   Why is the Johnson LMG not a true recoil operated gun?

              Answer   ....

                        ....

ANSWERS ON PAGE 185

9.

Safety Features

INTRODUCTION

Certain safety features must be included in the design of small arms to prevent damage to both weapon and firer. Firstly, there should be no possibility of the firing pin striking the cap until the round is safely supported by the chamber walls and the face of the bolt. Secondly, the cartridge case should be rigidly supported whilst the pressure is high. Thirdly, it should be possible to prevent the accidental operation of the trigger of a loaded weapon. Examples of some of the mechanical safety features found in small arms will be covered in this chapter starting with those which prevent the firing of round before it is safely chambered.

Mechanical Safety before Firing

The designer's main safety consideration for the period immediately preceding firing is to prevent cap strike occuring before the cartridge is safely held in the chamber. The cap is initiated when given a solid blow. This blow can come from a firing pin which itself needs to be driven forward by some external energy source such as a hammer. A typical hammer operated firing pin system is shown at Fig. 9.1.

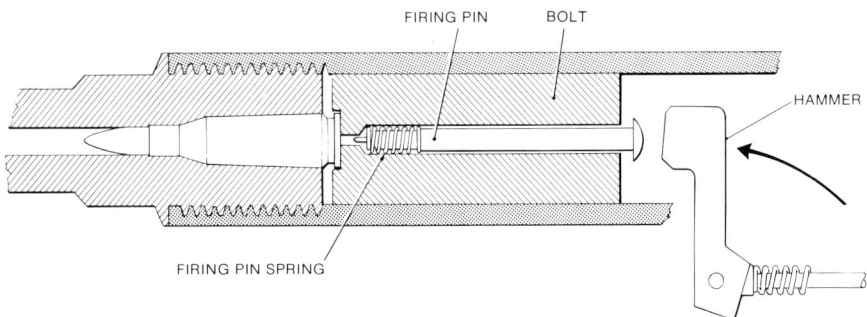

Fig 9.1 Hammer operated firing pin system

Alternatively the cap can be initiated directly by a striker. Unlike the firing pin, the striker has its own source of energy, often a spring, and in many weapons it is an extension of another component of the working parts. The diagram at Fig. 9.2 shows the striker on the US M60 GPMG which is an extension of the piston.

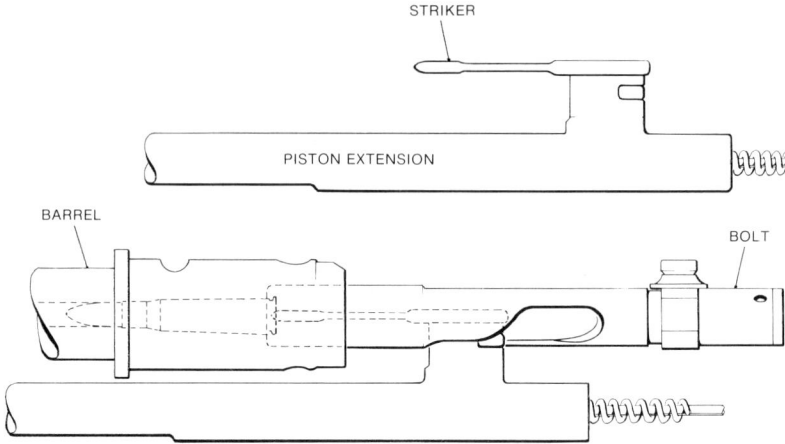

Fig 9.2 Striker on US M60 GPMG

The designer can adopt either one or a combination of measures to prevent the firing pin striking the round before it is properly chambered. Firstly, an obstruction can be placed in the path of the firing pin. Even if it is firmly struck the pin cannot reach the cap until the obstruction is removed. Figure 9.3 shows that rollers must move into recesses in the body of the West German G3 rifle before the firing pin can touch the cap. The rollers are prevented from moving into the recesses until the round is fully chambered.

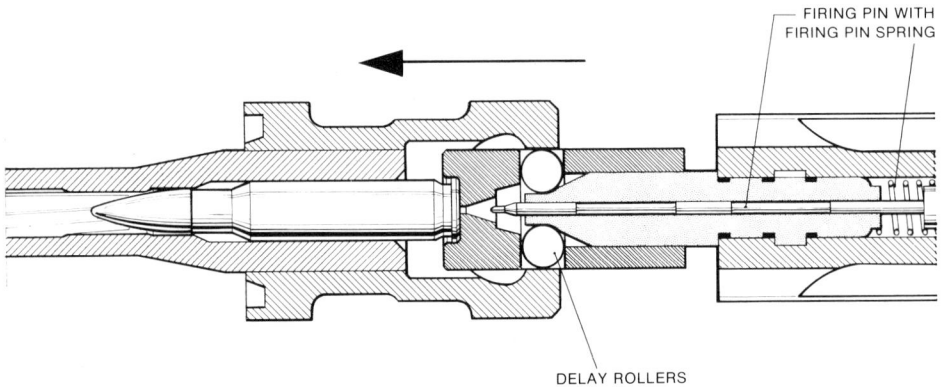

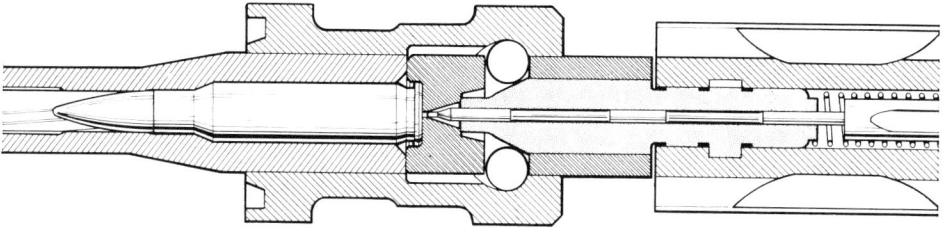

Fig 9.3   G3 Rifle Firing Pin Obstructed

Secondly, the axis of the firing pin is not aligned with the cap until the last minute.   The UK L2A3 is an example of a weapon which uses this method.

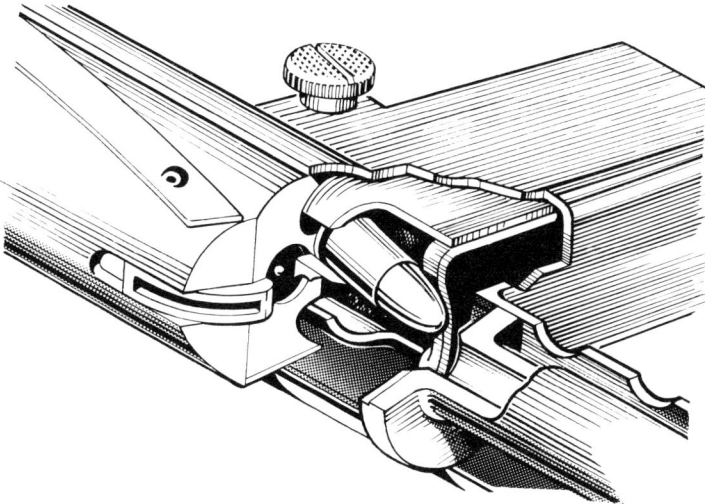

Fig 9.4 Non alignment of firing pin axis with the cap - UK L2A3 SMG

Thirdly, the hammer can be prevented from reaching the firing pin until the round is fully chambered. This third method alone is not completely satisfactory. As the bolt moves forward it is possible for the firing pin, without having been given any impetus by the hammer, to gain sufficient momentum to initiate the round when it strikes the cap. This danger can be prevented by using a light firing pin which is easily held back inside the bolt by a spring. The momentum which can be imparted to the firing pin when the weapon is swung violently downwards is an aspect of safety which has to be taken into consideration in order to avoid the danger of accidental discharge from closed bolt weapons.

Mechanical Safety after Firing

To achieve mechanical safety after firing the designer tries to ensure that the support for the empty case is not removed while the gas pressure is high. Under high pressure an unsupported case may be damaged and prevent the rapid firing of another round. In the worst instance the case may burst and the firer be at risk from hot debris and from hot high pressure gas escaping from the ejection opening. The designer has three options. Firstly, he can reduce the amount of high pressure gas, but low powered ammunition gives a low muzzle velocity. Secondly, he can limit the time the high pressure gas remains in the bore, but shortening the barrel so as to allow the gas to escape rapidly leads to low muzzle velocity and reduced accuracy. Thirdly, he can delay the rearward movement of the case until the pressure is at a safe level. This is easier to achieve with gas and recoil operated weapons than it is with weapons using the blowback principle.

In blowback systems the bolt is free to move as it is not locked to either the barrel or the body of the weapon. Simple blowback weapons rely on the inertia of the bolt and the strength of the return springs to prevent the bolt from moving back prematurely to leave the case unsupported at high pressure. To keep the weight of the bolt within reasonable limits simple blowback weapons must use low powered ammunition for safety. Advanced primer ignition systems permit the designer to use more powerful ammunition as the bore pressure is used to slow down, stop and reverse the movement of the bolt. However, as described in Chapter 8, special features are needed when high velocity rounds are fired by this system. Those on the Polsten HMG are illustrated in Fig. 8.7. Weapons which operate on the delayed blowback principle do not need any special safety features besides the delay mechanism. The delay mechanism can be comprised of rollers, as in the West German G3 rifle (Fig. 8.8) or of a simple lever, as in the French AA52 GPMG.

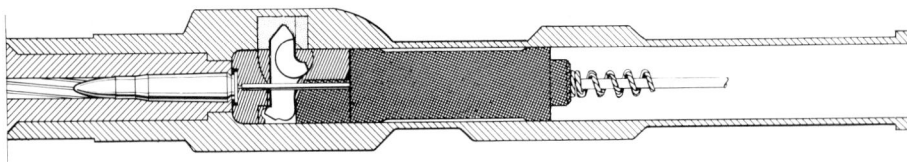

1   Position at Firing

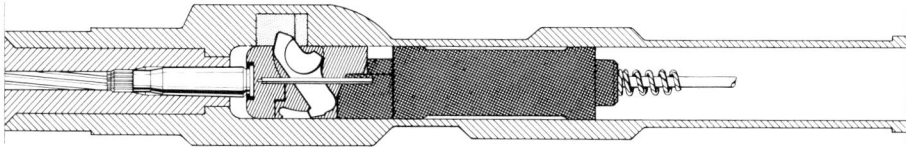

2   Lever rotating. Bolt body moving back

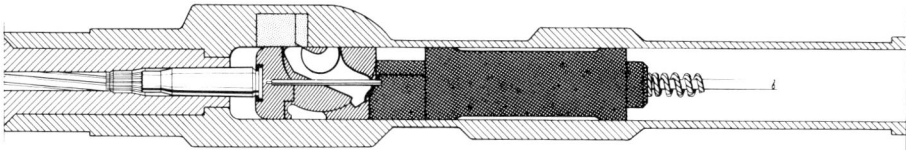

3   Lever disengaged from body. Entire bolt assembly moving back

Fig 9.5 Delayed Blowback Lever System. French AA52 GPMG

Even gas and recoil weapons require the empty case to leave the chamber while there is pressure in the bore so that an adequate rate of fire can be achieved and any mechanical linkages kept simple.   The moment of unlocking is decided by the need to use some blowback assistance or by the requirement for an acceptable rate of fire without excessive fouling or erosion.   In all systems there is a minimum delay between cap strike and unlocking which must take into account the time taken by the pressure in the bore to fall to a safe level.   For short recoil operated guns the minimum which the bolt and barrel move back locked together depends both on their inertia (their combined mass) and on the strength of the return springs.   For most weapons this is under 10 mm.   For gas operated weapons the necessary time delay is obtained by a combination of the time taken for the gas to operate the piston and the amount of free travel of the bolt carrier before unlocking occurs.   Figure 9.6 illustrates the delay mechanism of a gas operated weapon.

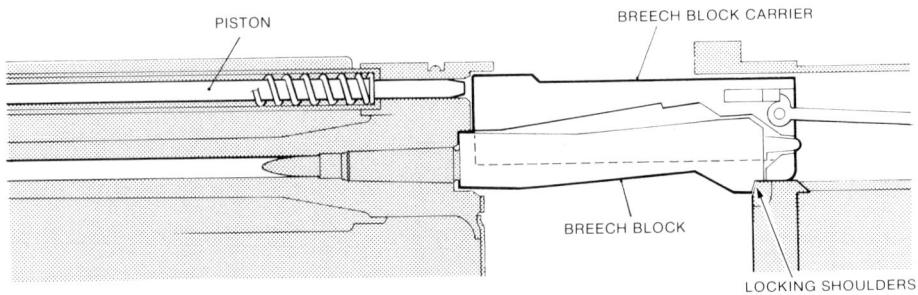

Breech block forward and locked

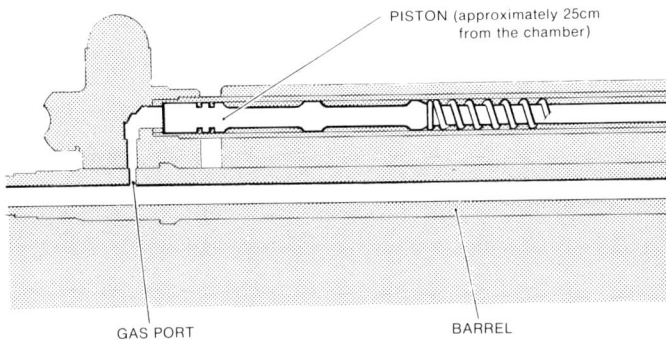

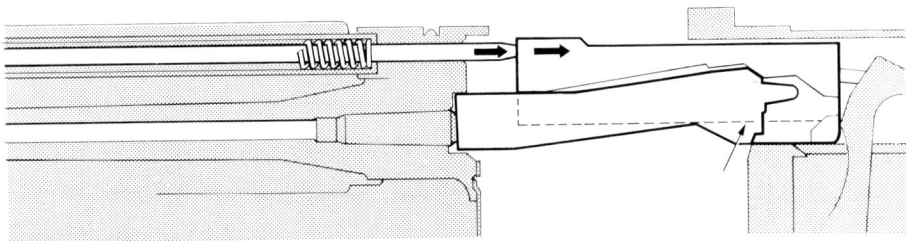

Piston pushing to rear (breech block unlocked by movement of ramps inside the carrier)

Fig 9.6 Delay Mechanism in a Gas Operated Weapon - UK L1A1

In the UK L1A1 SLR the gas cannot operate the piston until the bullet has passed the port. This takes just less than one millisecond. The pressure in the bore remains too high for safe extraction for a further 3-4 milliseconds. During this time the gas piston moves back to strike the bolt carrier. The force of this blow has to overcome the resistance of the return spring and the bolt carrier's inertia. As the bolt carrier moves backward it raises the bolt out of the locking shoulders but unlocking is not complete until the carrier has moved back approximately 12 mm. By this time the bore pressure has reached a safe level. This weapon uses a tilting block for the locking system. The shape of the locking

shoulders can be designed to assist in overcoming any primary extraction problems. The main disadvantage of this system is that much of the bolt and large
areas of the body are subjected to high firing stresses. These parts need to be
made of good quality metal which therefore increases the cost of the weapon.
One method of reducing the amount of high quality metal is the use of a forward
locking system. The most widespread one in use today with NATO and Soviet
weapons has a series of lugs on the front of the bolt as shown in Fig. 9.7.

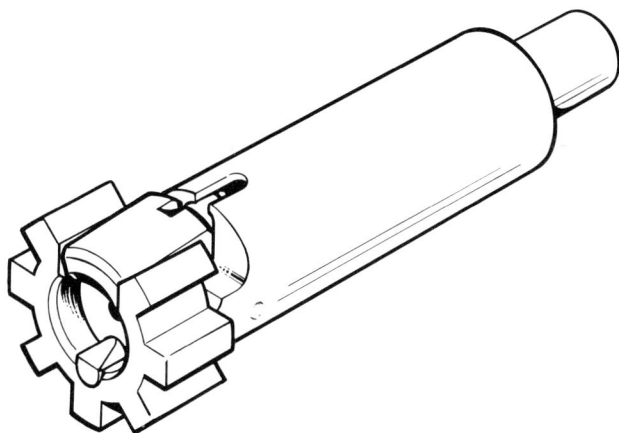

Fig 9.7 Front View of Typical Rotating Bolt

The bolt can be rotated by a curved cam path similar to the one in the US M60
shown in Fig. 9.8. The lugs then lock into recesses in the barrel.

BARREL   LOCKING LUGS   STRIKER   BOLT   STRIKER SPRING   FEED ROLLER

Before Locking

LUG LOCKED

PISTON EXTENSION

Locked

PISTON POST &
ROLLER

Unlocking

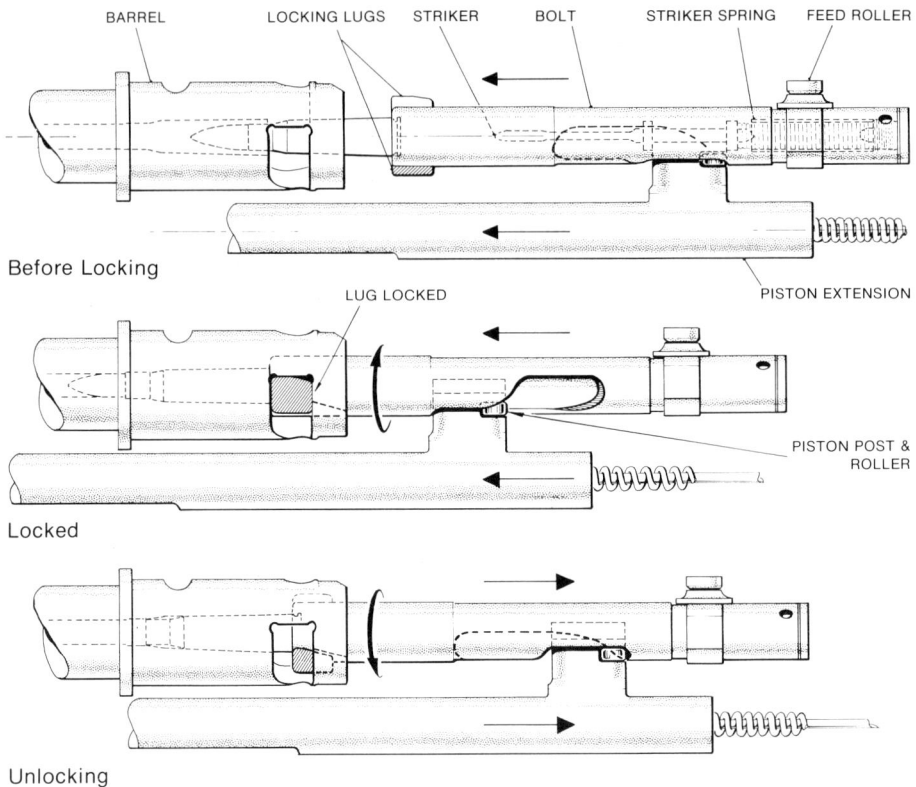

Fig 9.8 Rotating Bolt Locking System - US M60 GPMG

Applied Safety

Applied safety is the term used to describe measures designed to be operated by
the soldier to help him avoid an accidental discharge.   There are two methods of
achieving applied safety.   One method uses a safety catch which prevents the
firing mechanism from operating when applied.   The drawback of this method is
that the firer may neglect to use the safety catch.   Alternatively the firer can be
made to squeeze a release system in, for example, the pistol grip, before the
working parts are free to move.   Some weapons such as the Israeli Uzi SMG
combine both features.

The safety catch is often included in the fire selector switch.   Nearly all section
weapons have the facility to fire automatically in addition to single shots.   This
necessitates the design of a trigger mechanism which allows the working parts to
move back and forward only once, in other words single shot, and also allows
them to move continuously when firing bursts.   The trigger mechanism often includes more than one safety measure.   For example, it is possible to include

links within the mechanism to prevent the striker or hammer from moving until the round is chambered, even when the safety catch has not been applied. The safety catch itself either locks the trigger mechanism or disengages the trigger from the rest of the mechanism. Many of these features are well demonstrated on the Soviet AK - 47 rifle.

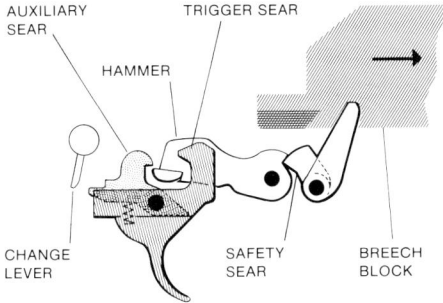

A   Hammer held on safety sear
    (Breech unlocked)

A. The AK47 fires from a closed breech. The diagram shows the breech block moving forward chambering a new round.
The safety sear prevents the hammer from moving upwards until the breech block is locked to the barrel.

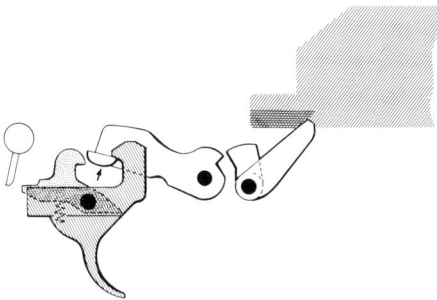

B   Hammer held on trigger sear
    (Breech locked)

B. The breech is now locked and the hammer is free to move when the trigger is operated.

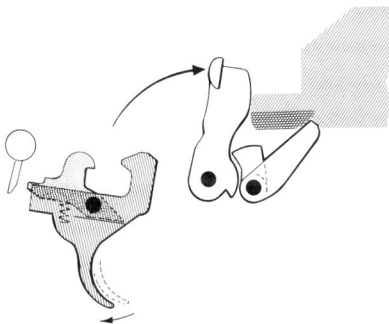

C   Trigger pressed; hammer released;
    gun fires

C. The change lever has been set at single shot. When the trigger is pressed the hammer swings up and hits the firing pin.

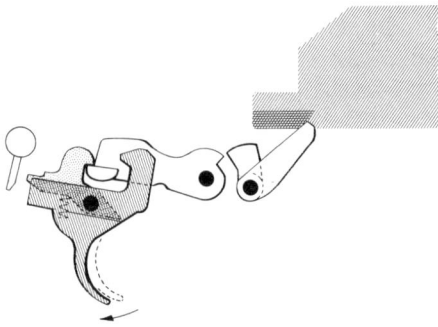

D   Hammer held on auxiliary sear

D. As the trigger mechanism is set at single shot, even if the firer keeps the trigger pressed, the weapon should not fire more than one round. The diagram shows the situation when the cycle of operations has been completed and a fresh round has been inserted in the chamber. Before this took place and whilst the breech was moving forward the hammer was again held by the safety sear until the round was chambered. The hammer cannot now move even after the safety sear releases it because it is held back by the auxiliary sear. This does not free the hammer until the trigger is released and the trigger mechanism returns to the position shown in diagram B.

Fig 9.9 AK47 Trigger Mechanism set to Single Shot

The change lever is now set at automatic. The weapon should continue to fire until the trigger is released. To achieve this the auxiliary sear should not restrain the hammer.

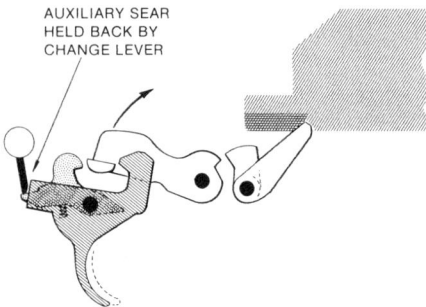

AUXILIARY SEAR HELD BACK BY CHANGE LEVER

The change lever holds the auxiliary sear clear of the hammer. Therefore when the breech is closed the hammer is free to swing up.

Breech forward and locked trigger pressed,gun fires

Fig 9.10 AK47 Trigger Mechanism set to Automatic

When the change lever is set at Safe the trigger is unable to move because the change lever itself becomes a bar across the back of the trigger sear. The trigger is therefore locked in position, as shown on the opposite page.

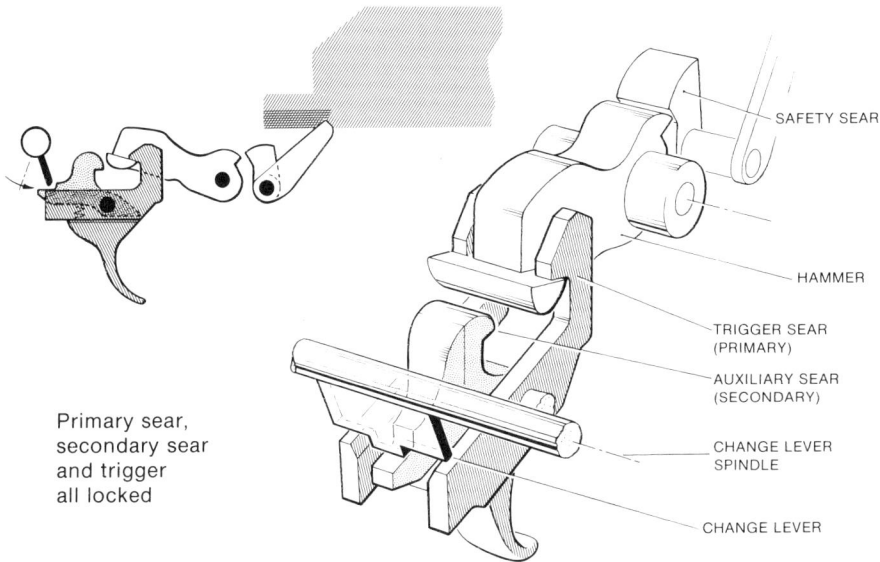

Fig 9.11 AK47 Trigger Mechanism when Safety Catch Applied

SAFETY SEAR

HAMMER

TRIGGER SEAR
(PRIMARY)

AUXILIARY SEAR
(SECONDARY)

CHANGE LEVER
SPINDLE

CHANGE LEVER

Primary sear,
secondary sear
and trigger
all locked

Small handy weapons such as pistols and SMGs are carried by soldiers whose primary role requires them to concentrate on matters other than handling their small arm.   Under these circumstances it is easy to forget to apply the safety catch.   To minimise the risk of accidental discharge from such weapons they are often fitted with a grip safety device.   The firer is required to hold the weapon firmly both in order to cock it and then to fire it.   The diagrams at Figs. 9.12 and 9.13 show these features in the Israeli Uzi SMG.   The weapon fires from the open bolt position.   Before it can be cocked the sear on the trigger must be withdrawn from the rear bent on the bolt.   This cannot be done unless the weapon is held properly and the grip safety released.

SEAR

REAR BENT

BOLT

FORWARD BENT

GRIP SAFETY STOP

PLUNGER AND SPRING

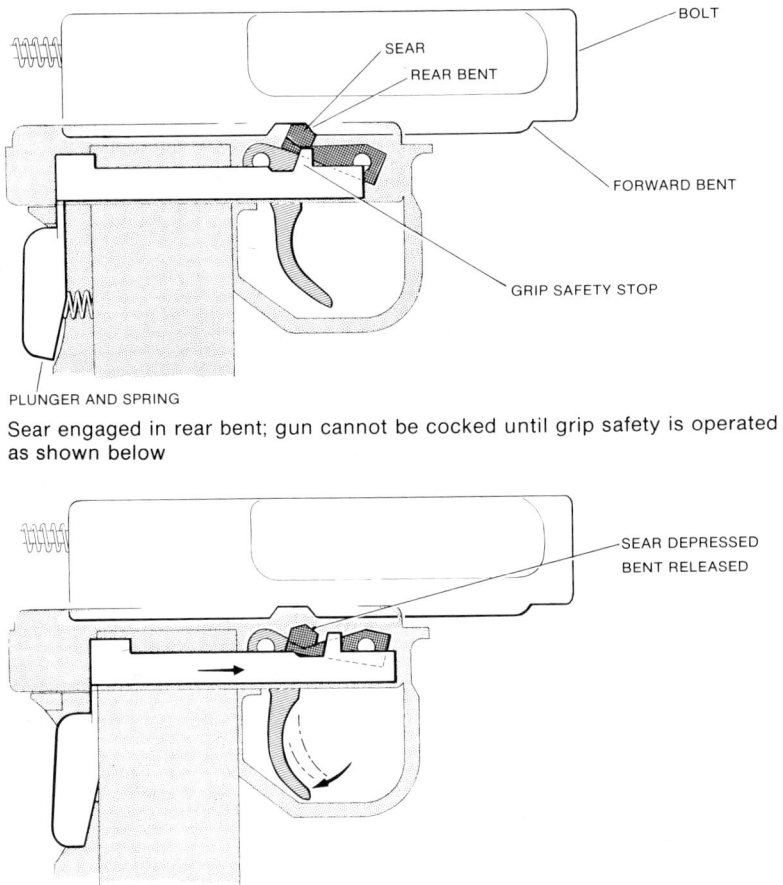

Sear engaged in rear bent; gun cannot be cocked until grip safety is operated as shown below

SEAR DEPRESSED

BENT RELEASED

Fig 9.12 UZI SMG Grip Safety preventing cocking

When the weapon is cocked the sear engages in the forward bent.   The trigger will not operate unless the sear is released by squeezing the grip safety device.

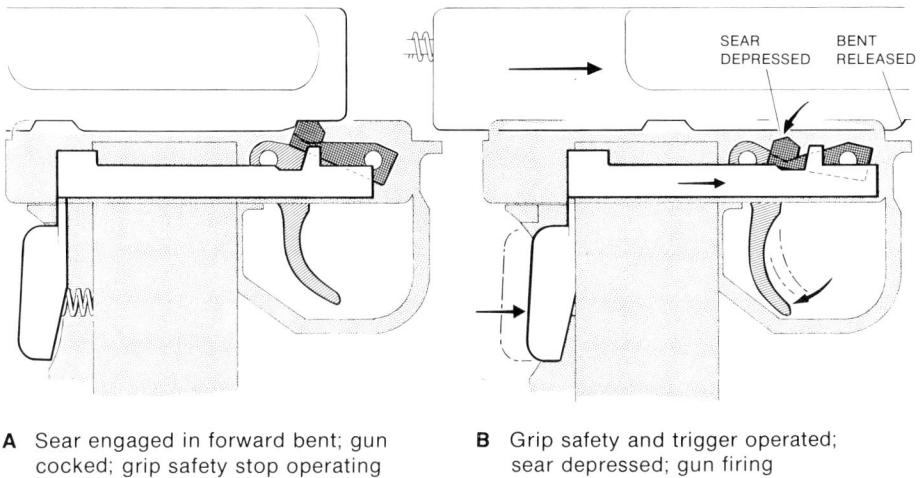

A Sear engaged in forward bent; gun cocked; grip safety stop operating

B Grip safety and trigger operated; sear depressed; gun firing

Fig 9.13 UZI SMG Grip Safety preventing the weapon

SUMMARY

Designers use two methods to ensure that weapons are mechanically safe. Firstly, the movement of the firing pin is controlled by the bolt so as to prevent the firing pin from reaching the cap until the round is properly chambered. Secondly, the round is kept fully supported after firing by preventing the mechanism which removes that support from operating while the gas pressure in the bore is high. Additionally all weapons are fitted with a safety device which either locks the trigger or disconnects the trigger from the rest of the mechanism. The firer is responsible for using this facility correctly to prevent accidental discharge. In delayed blowback weapons the delay mechanism controls both aspects of safety. In recoil operated weapons the locking device usually controls both the movement of the firing pin and the time interval after firing before unlocking occurs. In gas operated weapons which have a long stroke piston both safety aspects are controlled by the piston extension. In short stroke piston and direct action gas operated weapons the bolt controls the movement of the firing pin in conjunction with a safety sear, and the free movement of the bolt extension provides the necessary delay after firing.

SELF TEST QUESTIONS

QUESTION 1   What are the main safety considerations before and after firing?

Answer   . . . .

. . . .

QUESTION 2   Name two of the three methods which can be used by a designer to achieve mechanical safety before firing?

Answer   . . . .

. . . .

QUESTION 3   What methods are available for achieving mechanical safety after firing?

Answer   . . . .

. . . .

QUESTION 4   What is the purpose of a grip safety device?

Answer   . . . .

. . . .

QUESTION 5   How does the safety catch assist the firer in preventing accidental discharges?

Answer   . . . .

. . . .

ANSWERS ON PAGE 186

10.

Cannons

INTRODUCTION

A cannon is considered in some circles to be a small arm. The genre has grown out of the heavy machine gun (HMG) which was typically in the calibre range of 12.7 mm to 15 mm. In addition to being used to engage infantry at very long ranges, HMGs can be most effective against strong points and lightly armoured vehicles. Since vehicles and aircraft first appeared on the battlefield in any number (in World War I) they have grown ever more numerous, less vulnerable and faster moving. Their invulnerability in particular has progressed as armoured vehicles have proliferated. Not only the tank, but infantry combat vehicles and armoured reconnaissance vehicles have been enormously improved in design and armour, tending to make them much more difficult to defeat.

The result has been a dramatic increase in the calibre of the weapons needed to defeat these targets. The photograph at Fig. 10.1 shows a comparison of the sizes of 30 mm, 20 mm and 7.62 mm rounds. A 12 inch ruler is also included in the photograph.

Fig 10.1 30mm and 20mm rounds,
shown with a 7.62mm bullet

169

Aircraft too belie their flimsy forebears and appearance. Solid construction, the building in of redundant components and the addition of armour plate to protect vulnerable points means that a much larger, more destructive, possibly explosive projectile is necessary to be sure of inflicting the damage required for a kill. To attack both types of target has led to the development of special munitions like guided missiles, or greatly enlarged calibres in the sort of guns with which it was hoped to engage these targets. Cannon calibres have therefore developed and are commonly considered to be in the 20 to 35 mm range. Anything bigger, like the 40 mm Bofors, is already in a new class of specialised anti-aircraft weapons.

AMMUNITION

In the chapter on requirements we emphasised the importance of the ammunition. For cannon not only is the calibre an important aspect, but the type of ammunition which is feasible or desirable takes on a new look. To defeat the targets outlined above requires not only a bigger calibre but also a specialised round. The days when an overgrown bullet was adequate are past. To penetrate armours requires armour piercing ammunition, and to achieve the behind armour killing effect, or aircraft destruction may require an explosive projectile. The varieties of cannon ammunition currently in service include High Explosive (HE), Armour piercing (AP), Dual Purpose, Armour Piercing Special Effect (APSE), Composite and Armour Piercing Discarding Sabot (APDS). Since all of these are relatively expensive, purpose designed cheaper training rounds are also available.

High Explosive

HE ammunition is no more nor less than a small shell, bringing with it several complications. The first is to get an adequate explosive charge into the small volume and still have a shell wall strong enough to survive the stresses in the chamber and barrel. Special techniques and materials may be necessary. A fuze is needed which has to work almost infallibly despite the necessary miniaturisation. Since usually this round will be employed in the anti-aircraft role, live shells must not fall among friendly troops, and a self destruct mechanism is needed. This sort of round is now of little use against modern armoured vehicles but it still has a purpose against semi hard targets, aircraft, soft skinned vehicles, buildings and weapon emplacements.

Armour Piercing

AP ammunition may look very like an enlarged small arm round, but is likely to have improvements in design and materials to assist penetration. Charge weight may be increased to enhance muzzle velocity. This brings two benefits: a flatter trajectory leading to greater accuracy and improved first round hit chance; and greater energy for improved penetration. Simple AP rounds may lose effect because they pass through both sides of an AFV, causing neither damage nor casualties on route. Most other designs of round seek to overcome this potential defect.

Dual Purpose

This type of round seeks a compromise between the two main potential targets.
It has some armour piercing capability if required but if an aircraft or soft skin
vehicle is engaged there is also an HE effect. This brings a bonus in that if any
armour is penetrated there is then an explosive behind armour effect. The
Norwegians have done a lot of work to market such a round, which though not
optimised for either task gives a good capability spanning both roles.

Armour Piercing Special Effect

APSE is a UK development aimed at achieving the behind armour effect which will
result in a tank kill. This round, once it has penetrated produces searing hot
flames within the vehicle, known as the pyrophoric effect.

Composite

Composite armour piercing rounds are designed to improve penetration. This is
done by surrounding a small solid core shot by a light steel or alloy body of
similar outline to a conventional round. The remaining energy at the target is
then concentrated in a smaller diameter so that thicker armour can be defeated
than with a full calibre round. A bonus effect is that the alloy casing gives a
flash effect on the armour, acting as a marker for the gunner who may not otherwise be sure that he has hit.

Armour Piercing Discarding Sabot

Cannons are required to defeat the most heavily protected aspects of light AFVs.
The slopes, thickness and improved designs of armours, even on APCs, makes
this a formidable task. From a gun, at present, this can only be achieved by
APDS. The necessary levels of penetration cannot be obtained from hollow charge
ammunition which is small enough to fit the bore of a cannon. Although apparently
a scaled down version of tank gun APDS, designers have discovered that some
aspects of cannon APDS need to be redesigned in order that it is effective in all
respects. APDS is a sub calibre shot surrounded by light alloy or plastic petals
and driving band which are discarded at the muzzle of the gun. The highly aerodynamic shot has a muzzle velocity of a very high order which it retains for a
considerable range. This is an advantage over composite shot which loses
velocity more rapidly. Again the remaining energy acts against a small diameter
to give good penetration, assisted by the dense nature of the shot material, which
usually incorporates tungsten.

Ammunition Trends

Ways to improve the effectiveness of cannon rounds are being actively researched.

In the anti-armour field the need for high specific weight and a behind armour
effect has been stressed. A shot core material conferring both is Depleted

Uranium (DU). This is naturally very heavy as uranium has such a high atomic number, and on piercing armour it has a most spectacular and instantaneous pyrophoric effect. Some countries, unworried by the environmentalist lobby, are progressing towards production of DU ammunition. Whilst radioactivity levels are so low as to present no hazard, other Governments are dithering, afraid to risk an uninformed furore from those who set out to embarrass the defence community on any issue. As an example of the safety of the material, it has been used for years as counterweights to control surfaces in civil aircraft wings, and to add weight to the keels of yachts.

For the anti-aircraft role the HE and dual purpose rounds are fairly effective. All types of AP rounds would have some effect though they are an expensive item to use in large numbers. This is necessarily the case when, in order to hit high speed aircraft, one has to put as many rounds as possible into the path of the aircraft in the hope that some at least will meet their target. The developments being tried to improve the chance of a hit include increasing the velocity of the rounds and of course dramatically increasing the cyclic rate of fire, not least by the use of multi barrelled weapons. A further improvement to the ammunition is to include more than one projectile in each round. One design with some promise is a warhead containing three ring aerofoils. Each is slightly aerodynamically different so that they separate, and sweep three times the volume of sky. Their drawback at present is that individually they are not very lethal.

Cannon rounds are not small, and the trend is to the bigger calibres in order to defeat the required targets. Most cannon are mounted on aircraft or vehicles in which space is at a premium. The incentive to seek smaller ammunition by designing a caseless or liquid propelled round is therefore as high if not higher than with small arms. Several countries are working on the problem. The engineering and environmental difficulties facing a caseless cannon round should be less than with a weapon which needs to be carried and fired by a man, and thus not exposed to the full range of environmental hazards. It is highly likely therefore that a cannon with unconventional ammunition, probably caseless, will come into service sometime before the end of the 20th century.

CANNON EMPLOYMENT

As mentioned previously, the targets for which cannon have developed are largely mechanical. Aircraft, vehicles with and without armour, and small maritime vessels are the main categories. Not surprisingly cannons have therefore been adapted for mounting on the same variety of systems. The anti-aircraft role has led to multiple mountings in order to put a heavy concentration of projectiles into the path of an aircraft.

The photograph opposite at Fig. 10.2 shows the triple mounting used on a Yugoslavian anti-aircraft cannon.

Fig 10.2 A typical AA mounting

These mountings have also included some elaborate sighting systems and remote firing devices. Radars are often used to detect targets, and in association with computers, control the firing of the guns. The weapons themselves, however, have been fairly conventional for a long time and also common to each role. An excellent example is the 20 mm Oerlikon, a cheap but effective weapon of World War II and after, which was to be found operating on land, sea and in the air.

The cannons of the last 30 years have been reliable and effective, so that not much change in their traditional design and operation has occurred until very recently. The traditional designs were really only big brothers of the machine gun. They may have been blow back or gas operated guns with minor modifications as found necessary, such as advanced primer ignition and blow back with locked breech. There were complications in feed mechanisms because of the weight of the rounds, and recoil mechanisms were often incorporated since the guns were mounted in very light vehicles. The trend away from these traditional style weapons has been brought about by the conflicting demands of meeting more than one battlefield role and of defeating the ever more difficult targets. The extremes of target are the modern high performance aircraft and the well armoured vehicle.

To give a good chance of a hit on a high speed aircraft the rate of fire must be as rapid as practicable. Some cannon of traditional design have, with ingenious modification, been capable of producing up to 1,000 rounds per minute or more; but the scope for further improvement is limited and additional complications such as reduced reliability occur. The Vietnam conflict led the USA to develop for land service use, cannons which previously had only been mounted in aircraft. Apart from the anti-aircraft role, the need adequately to neutralise large areas of jungle and delta country made them adopt the 'spray the area' concept of covering fire. The externally powered gun is able to achieve a rate of fire which is simply varied by the application of more or less power. Thus it could be quickly adjusted to cater for the high rates demanded by the roles mentioned above, or toned down to take on the point target. Even then rates could not be made adequate for all purposes, and the solution was to develop multi barrelled versions. Using the old Gatling principle of rotating a number of barrels and bolts, rates of fire of up to 6,000 rounds per minute can be achieved.

External power is needed not only to rotate the barrel but also to feed the rounds. Originally developed for helicopters, the principal has now been adapted to vehicles. The General Electric Company produced the 20 mm Vulcan and the 7.62 mm Mini gun both with 6 barrels, and a "lightweight" 3 barrelled 20 mm version. A feature of the externally powered gun is its far greater reliability than the conventional weapon, primarily because the round is under positive control throughout its cycle. Dispensing with gas for any aspect of the cycle of operations allows the toxic fumes to be channelled away from the interior of the vehicle in which it is mounted, and the empty cases too can be ejected outside the vehicle. All these advantages to externally powered weapons has led to others being introduced. The Hughes armaments company of California produced their chain gun, so called because a chain like that of a motor cycle drives the gun's working parts. The 30 mm chain gun developed for helicopter mounting has now been adapted to 25 mm and chosen for the US Bushmaster system, which will be mounted on their infantry and cavalry combat vehicles. The adaptability of design is illustrated by the weapon now being available in 7.62 mm calibre and it is a likely choice for coaxial mounting on British AFVs.

At the other extreme to the gun offering high rates of fire is the gun optimised to take on point targets. The firing of a round creates vibrations in the barrel which reduce the accuracy of subsequent rounds if they are fired before the vibrations have ceased. To be certain of hitting a target the size of an APC at 1,000 metres virtually demands single shots. The UK decided that this was to be the priority task, and only sought a deterrent effect against aircraft. The outcome was the design at the Royal Armament Research and Development Establishment, together with the Royal Small Arms Factory, of the 30 mm Rarden gun. This fires from a ready clip of 6 rounds only on single shot, which can be repeated very quickly, but careful design ensures that the vibrations settle rapidly between shots. The result is an extremely accurate gun, which taken in conjunction with its quite large calibre and specialised ammunition gives it a commendable chance of killing light AFVs. Figure 10.3 opposite shows the Rarden 30 mm cannon mounted on Fox reconnaissance vehicle.

**Fig 10.3 RARDEN 30mm gun mounted
on Fox combat vehicle reconnaissance**

A further factor tending to deter some military requirements staffs from asking
for guns with high rates of fire is that they inevitably get through a huge quantity
of ammunition which has to be stored somewhere on the vehicle. If the vehicle
is principally wanted to carry men and their equipment round the battlefield,
cannon ammunition has to compete for the limited space available. In the
British philosophy, going for smaller rounds would be pointless, since lethality
is of paramount importance.

To answer in part the conflicting demands of potential targets, there has been
another interesting surge in the development of alternative feeds. It is awkward
to find when faced with an armoured target, that the ready round is one which has
been developed for the anti-aircraft role. To cater for this some nations have
introduced dual feeds, with the facility to switch from one to another as the
occasion demands. West Germany's Rheinmetal Mk 20 Rh202 20 mm has this
facility, which is shown overleaf in Fig. 10.4.

Fig 10.4   Alternative Feed Mechanisms - The Rheinmetall Mk20 Rh202

FUTURE TRENDS

The Americans have in development a modular cannon, adaptable to several calibres, and called the Dover Devil. It is specifically designed to tackle any potential target, men, aircraft and vehicles. It has a dual feed which can readily switch ammunition to match the target, and yet a third feed which can be brought nearly as quickly into action.

The overall picture then of cannon today is of trends towards specialisation, for either high rates of fire or extreme accuracy; to ease of mounting and small recoil forces; to externally powered guns of high reliability; and to purpose built rounds of ammunition which complex feed arrangements can make available for the appropriate target.

SELF TEST QUESTIONS

QUESTION 1   What are the conflicting requirements of cannon, leading
              usually to a compromise in design?

              Answer   ....

                        ....

QUESTION 2   What problems face ammunition designers in the defeat of armour
              by cannon?

              Answer   ....

                        ....

QUESTION 3   What future ammunition development is most likely and why?

              Answer   ....

                        ....

QUESTION 4   How can the chance of hit on aircraft be improved now and in
              the future?

              Answer   ....

                        ....

QUESTION 5   List some varieties of cannon ammunition now in service.

              Answer   ....

                        ....

                        ....

QUESTION 6   Outline the British philosophy leading to the Rarden design.

              Answer   ....

                        ....

QUESTION 7   What is the chief role of modern cannon?

              Answer   ....

QUESTION 8   Why is HE ammunition for cannon a problem to manufacture?

              Answer   ....

                        ....

QUESTION 9   What does multi-purpose ammunition set out to do?

Answer   . . . .

. . . .

QUESTION 10   What is the alternative to dual purpose ammunition when faced
with uncertainty as to target type?

Answer   . . . .

. . . .

ANSWERS ON PAGE 186

Answers to Self Test Questions

CHAPTER 1

Page 16

QUESTION 1   The person to person encounter. Flushing out defenders. Anti-terrorist encounters. The boost to morale from a personal weapon. Final covering fire in an infantry assault. Specialist roles such as sniper.

QUESTION 2   That 95% of all rifle engagements take place at 400m or less.

QUESTION 3   A lighter load for the infantryman. Reduction in the effective range expected of small arms. Reduction of logistic load. Handleability (especially in vehicles). Improved shooting from the average soldier. Awareness that incapacitation is possible by ammunition design.

QUESTION 4   Rate of fire. Accuracy. Consistency. Penetration. Incapacitating effect. Effective range. Weight. Length. Portability. Reliability. Ease of maintenance. Robustness. Simplicity of operation. Ease of training. Secondary roles.

QUESTION 5   Primarily the form of ammunition presentation (magazine for LMG, belt for MMG), and the mounting (bipod for LMG, tripod for MMG). Also operating range   1000m for LMG   2000m for MMG.

QUESTION 6   Danger from toxic fumes. Likelihood of stoppages. Too much inboard space taken up. Poor barrel changing capability.

QUESTION 7   Infantryman's rough handling. Operation in heat, cold, wet and dirt. Operation by troops in special clothing (NBC suits and arctic gloves).

QUESTION 8   Particularly by showing that incapacitation can be achieved by a bullet optimised on one parameter (calibre/mass, velocity or stability).

QUESTION 9   Inadequate performance of overgrown conventional bullets against
               mechanical targets (AFVs and aircraft) and obsolescence in the
               anti-personnel role.

QUESTION 10   Different emphasis on specific characteristics. Historical lessons
               remembered. Designer's quirks. Military priorities. Commer-
               cial considerations.

CHAPTER 2

Page 30

QUESTION 1   Mass, shape, size and ductility (ability to fragment).

QUESTION 2   Render the target incapable of carrying out its task.

QUESTION 3   a.   The amount of energy which is transferred.
               b.   The rate of energy transfer.
               c.   The location of the wound.
               d.   The motivation of the target.
               e.   Whether or not the projectile is moving at a high velocity.

QUESTION 4   Creation of a shock wave in the target material to extend the area
               of damage beyond the bullet track.

QUESTION 5   a.   Increased flash.
               b.   A small reduction in muzzle velocity.
               c.   Increased recoil as loss of weight more significant than the
                     loss in muzzle velocity.

QUESTION 6   a.   Rough handling by soldiers.
               b.   Grenade projection and bayonet fighting.
               c.   Heat dissipation (amount of heat depends on the rate of fire).
               d.   Pressure produced by the burning propellant.

QUESTION 7   a.   An increase in the spin of the bullet which affects its stability.
               b.   Wear rate could be affected. If increased too much the bullet
                     jacket could become overstressed.

QUESTION 8   a.   The commencement of rifling.
               b.   This is the place of maximum stress on the lands and is a
                     likely place for gas wash erosion.

QUESTION 9   The premature ignition of the propellant due to heat.

QUESTION 10   The chamber walls become hot enough to cause cook off far
               quicker than with brass cased ammunition.

CHAPTER 3

Page 51

QUESTION 1
a. Strike energy in terms of $\frac{1}{2} mv^2$ ($mv^3/2$ more strictly).
b. Ability to transfer more than 80 Joules.
c. Stability of the bullet - more energy is transferred if the cross sectional area can be increased, for example by tumbling.

QUESTION 2
a. Stability - the greater the stability the greater the penetration.
b. Strike energy ($\frac{1}{2} mv^2$).
c. Density of the material which is being penetrated.

QUESTION 3
Against unprotected men 2000m. This can be reduced by over half if the man is wearing a steel helmet or some form of body armour.

QUESTION 4
a. Greater range as its carrying power is greater.
b. Gains more energy from the propellant when in the bore.
c. Deflected less by cross winds.

QUESTION 5
a. Leads to increased recoil of the weapon.
b. Needs greater kinetic energy to penetrate any material.

QUESTION 6
a. The bullet could shatter on impact with hard materials such as brickwork or thin armour.
b. The length/diameter ratio must be less than 5 for stability reasons, so if a tracer round is needed in the same calibre, it will be excessively long.
c. It may be necessary to increase the spin rate to maintain stability; this could create wear problems.

QUESTION 7
a. The advantages are handiness in a confined space and some weight reduction.
b. The disadvantages are flash, inefficient use made of the propellant (loss of muzzle velocity), volume of fire may be restricted because of overheating and recoil may increase.

QUESTION 8
a. Trying to match the tracer trajectory to that of the ball round whose centre of gravity and mass remain constant throughout the flight.
b. Ignition of the tracer composition.
c. Seeing the tracer in daylight.
d. The amount of composition that can be fitted in the round, its burning rate and so the effective range of the tracer.
e. Stability problems associated with keeping the l/d ratio below 5.

QUESTION 9 a. Enemy soldier, what will he be wearing, how quickly should
                he be stopped from continuing with his task.
            b. What strike energy does the bullet need to achieve the above
                at the tactically desired range.
            c. What muzzle velocity is required to achieve that strike energy
                at that range and also gives the bullet the desired vertex
                height and the acceptable wind deflection.
            d. What charge weight is needed to produce that muzzle velocity.
            e. To keep the recoil acceptable will the weapon weight also be
                acceptable to the soldier.

QUESTION 10 a. Range: 2000m for MMG role; perhaps as close as 600m for
                the LMG.
            b. Volume to Fire: A much higher rate is expected of an MMG
                which can only be met by barrel changing or water cooling.
            c. Weight: Lightness is more important in an LMG.

CHAPTER 4

Page 75

QUESTION 1   The rate of heat input from the ammunition far exceeds the ability
             of a small arm to lose heat to its surroundings even at modest
             battlefield rates of fire.

QUESTION 2 a. The ammunition cooks off.
            b. The weapon becomes too hot to hold.
            c. The fire becomes inaccurate because of excessive wear and
                erosion.

QUESTION 3   Radiation heat loss depends on the fourth power of the absolute
             temperature, whereas conduction and convection losses depend
             directly on the temperature. The difference between T^4 and T
             becomes more pronounced as the temperature rises.

QUESTION 4   The rate of heat loss from the barrel surface is far slower than
             the rate of conduction of heat from the bore to the surface. To
             keep this build up of heat within manageable proportions it is
             necessary to slow down the rate of heat flow from the bore. This
             can be done by increasing the thickness of the barrel walls.

QUESTION 5   The round is most unlikely to cook off unless this temperature is
             maintained for many minutes. The chamber is likely under such
             circumstances to reach around 180-200°C which is the ignition
             temperature range of small arms propellants. The brass car-
             tridge case however may well provide sufficient insulation.

QUESTION 6 a. Wear is the gradual removal of layers of metal from the bore
                and is caused by the bullet rubbing against the bore.

b.   Erosion is the removal of particles of the bore and is the damage caused by high temperature gases at high pressures.

QUESTION 7   Gas wash erosion occurs when hot high pressure gas is forced through a narrow aperture. The bore surface may well melt.

QUESTION 8   If forced convection is used and if the fin spacing minimises mutual heating by radiation.

QUESTION 9   Changing barrels which have chromium plated bores or other hard liners. Other methods are likely to produce weapons which are too heavy.

QUESTION 10   Keeping the enemy's head down and making movement hazardous. It can be done by firing at least 60 rds/min to within about 15m of an enemy position.

CHAPTER 5

Page 95

QUESTION 1   The average dispersion of a group of shots around an MPI. Lowest acceptable FOM for UK 7.62 mm ball is 8 inches at 500 yards.

QUESTION 2   High rates of fire, with heat as an incidental contribution. Very small calibres (under 5 mm). Increasing the twists of rifling for the length of barrel.

QUESTION 3   A muzzle will droop due to the softening of the barrel metal and result in MPIs falling lower on the target.

QUESTION 4   Any sighting error will be minimised because the angle of error is smaller.

QUESTION 5   Muzzle velocity remaining constant results in consistency of trajectory, and therefore improved accuracy. It is achieved by keeping chamber pressure as constant as possible.

QUESTION 6   A high MV results in quicker bullet exit and a flat trajectory. A lower MV results in slower bullet exit and a higher trajectory.

QUESTION 7   The weapon. The ammunition. The firer.

QUESTION 8   The centre of gravity of the weapon lying off the axis of the bore. The point of support in the firer's shoulder lying off the line of recoil.

QUESTION 9   Accurate relocation. Rigid support. Optimum offset and height for the firer's eye. A facility to zero.

QUESTION 10   The engagement of individual personnel targets in either a major war or internal security scenario. Making shots count as the number of small arms available to fire falls. To give confidence to the user.

CHAPTER 6

Page 106

QUESTION 1   One which is integral to weapon performance, and one which offers a secondary role.

QUESTION 2   Reflected blast, harmful to the firer's ears.

QUESTION 3   Flash and noise (silencers).

QUESTION 4   By deflecting the gases at the muzzle upwards, drive the muzzle down, thus reducing the tendency of automatic weapons to climb.

QUESTION 5   Extra weight to carry. Weapon not employed in primary role. May need special round.

QUESTION 6   By giving the grenade a tail and fins, and a sight on the spigot.

QUESTION 7   By live bullets either passing through the grenade and the following gases propelling the grenade, or by a bullet trap in the grenade.

QUESTION 8   To break up any projectile which is needed to give a semblance of the normal cycle of operation with a training round.

QUESTION 9   Either a cone, or a series of bars projecting from the muzzle.

QUESTION 10   Type of warfare is less pertinent. Small light weapons are not suitable for fixing bayonet.

CHAPTER 7

Page 131

QUESTION 1   Whether portability and rapid reloading are more important than a sustained rate of fire.

QUESTION 2   For: lowers silhouette, is out of sight line. Against: difficult to change, spring must overcome gravity, a long one interferes with the ground and limits depression.

QUESTION 3   Box, drum and tubular.

QUESTION 4   Continuous and disintegrating link.

QUESTION 5   Indicates when the magazine needs replacing. Speeds replacement. Avoids the danger of cook off. Weapon cools more quickly.

QUESTION 6   Gun driven. Spring driven. Externally driven.

QUESTION 7   Fixed, rocking and push rod.

QUESTION 8   The bent is a recess in the firing mechanism. The sear is a projection in the trigger mechanism. Sear engages in bent to hold firing mechanism in cocked position. Disengagement starts firing process.

QUESTION 9   Reloading is quicker and easier. Heat dissipates more quickly and cook off is avoided. Accuracy is marginally less important.

QUESTION 10   Extraction and ejection.

CHAPTER 8

Page 152

QUESTION 1   a.   It is the distance between the bolt face and a specified point on the cartridge. This point is not the same for rimmed and rimless rounds.
              b.   Excess cartridge head space may either prevent the gun from firing as the safety mechanism may stop the hammer being released or the base of the case may be unsupported which could lead to rupture at the neck or the side near the base.

QUESTION 2   Blowback weapons have bolts which are not locked to either the body or the barrel. However note hybrid systems such as blowback with locked breech.

QUESTION 3   Flutes for lubrication, so as to make primary extraction of an empty case easy.

QUESTION 4   They have relatively heavy bolts which on pressing the trigger move forward from the open bolt position. Lock time is long, the centre of gravity of the weapon changes and so the point of aim is likely to wander.

QUESTION 5   Unlike the short recoil operated weapons, in those with a long recoil no part of the cycle of operations can take place until the barrel has started to move forward again.

QUESTION 6   Lack of power at small calibres (7.62 mm and below), because there is not enough propellant in the case (less than 1% of the propellant energy is converted to recoil).

QUESTION 7   a.   Erosion.
               b.   Carbon fouling.
               c.   Speed of unlocking.

QUESTION 8   The gun slows down and empty cases may not be ejected cleanly.

QUESTION 9   Recoil - the others vent too much gas into the crew compartment.

QUESTION 10   The energy from the recoil forces only unlocks the breech and is insufficient to power the cycle of operations. The energy for the latter comes from blowback action.

CHAPTER 9

Page 168

QUESTION 1   a.   Before Firing: to prevent cap strike occurring before the cartridge is safely held in the chamber. Also that the inadvertent operation of the trigger will not fire the round.
               b.   After Firing: support for the empty case is not removed while the gas pressure is high.

QUESTION 2   a.   Firing pin or striker is obstructed.
               b.   Axis of the firing pin or striker is not aligned with the cap.
               c.   Hammer is prevented from reaching the firing pin.

QUESTION 3   a.   Reduce the amount of high pressure gas. (Use low power ammunition).
               b.   Limit the time that the gas pressure is high. (Shorten the barrel).
               c.   Delay the rearward movement of the case until the pressure is low.

QUESTION4   To minimise the risk of accidental discharges by forcing the firer to hold the weapon firmly as though about to fire it.

QUESTION 5   If applied, the firing mechanism cannot operate.

CHAPTER 10

Page 177

QUESTION 1   A high rate of fire against aircraft and accuracy against vehicles.

QUESTION 2   Increasingly tough armour. Possibility of going through both sides without impairing fighting efficiency. Need for behind armour effect.

QUESTION 3   Caseless rounds since storage space is limited in cannon mounting weapon systems, and to reduce ammunition costs.

QUESTION 4   High cyclic rates of fire. Multiple barrels. Multiple projectiles. Associated detection and direction equipment.

QUESTION 5   HE, AP, APSE, APDS, Dual purpose, training.

QUESTION 6   Lethality first (why fire if no kill?) therefore large calibre. To kill you must hit, therefore high accuracy, ie not automatic.

QUESTION 7   Destruction of point targets, most often vehicular (land, sea or air).

QUESTION 8   The small size imposes limits on the HE content, on shell wall thickness, and on the space to include efficient fusing.

QUESTION 9   To provide a ready answer to whichever target is presented, requiring HE and/or AP.

QUESTION 10 Multiple feeds from which the appropriate round can quickly be selected.

Glossary

A

Ablative propellant
   Propellant containing chemicals which reduce wear and erosion.

Absolute temperature
   Temperature scale used in calculating the amount of radiated heat.
   0° Absolute = -273^OC.

AFV
   Armoured fighting vehicle.

Annular
   Ringlike.

AP
   Armour piercing.

APC
   Armoured personnel carrier.

APDS
   Armour piercing discarding sabot.

Apogee
   Highest point on the path taken by a bullet.

APSE
   Armour piercing special effect.

Assault rifle
   Infantryman's personal weapon of the SMG type with improved
   characteristics of range and accuracy, though not equal to the traditional rifle's performance in these parameters.

Automatic
   The firing of a succession of rounds until the trigger is released or
   the ammunition runs out.

B

Ball round

Normal lead core jacketted ammunition.

Beaten zone

That horizontal area of ground onto which rounds in a burst from an automatic weapon fall at the end of their trajectory.

Bent

A recess into which fits the sear, a component of the trigger. When these two engage the part containing the bent is prevented from moving, usually forwards.

Bore

Central hollow running the length of the gun barrel.

Breech

The pressure resistant metal casing surrounding the chamber.

Breech block or bolt

The component which holds the base of the round while it is being chambered, fired and extracted.

C

Calibre

The nominal internal diameter of the bore or the nominal external diameter of the projectile.

Cannon

Weapons of 20-40 mm calibre.

Cap

Portion of the base of the cartridge case which contains the primer composition.

Carbine

A shortened version of the traditional rifle design, originally for mounted troops, later favoured in close terrain such as jungle.

Caseless ammunition

Ammunition which does not have a metal case to contain the propellant and provide rear obturation.

Chamber

That part of the bore into which the cartridge case is loaded.

Charge mass

The amount of propellant contained in the cartridge case.

Closed bolt

    Weapons which fire from a closed bolt have a round in the chamber before the trigger is pressed.

Cocking

    The process of making the weapon ready to fire.

Cook off

    The unintentional firing of a round owing to the propellant becoming too hot.

Cyclic rate

    The number of rounds per minute which an automatic weapon can continuously fire, assuming an unlimited supply of ammunition and no stoppages.

E

Ejection

    Expelling of an empty case from the body of a weapon.

Emissivity

    A measure of an object's ability to absorb and radiate heat.

Erosion

    Removal of particles of the bore by the hot propellant gases.

Extraction

    The withdrawal of an empty case from the chamber.

F

FOM (Figure of Merit)

    A measure of the dispersion of a representative sample of a batch of ammunition.

Furniture

    Weapon parts whose only function is to give improved handling, eg cheekrest and handguard.

G

GPMG (General purpose machine gun)

    An automatic weapon capable of being adapted to either the light or medium machine gun roles.

Group

    A number of rounds fired at the same point of aim on a single target.

H

Hand gun

Pistol, revolver or 'automatic' capable of being fired from one hand.

HE

High explosive.

HMG (Heavy machine gun)

An automatic weapon of calibre greater than $\frac{1}{2}$ in (12.75 mm) but of less than the 20 mm cannon calibre.

I

ICV

Infantry Combat Vehicle

Impulse

The impulse given to an object is mathematically equal to the force on that object multiplied by the time during which the force is applied. (Newton second).

IW (Individual Weapon)

It is similar in performance to an assault rifle.

J

Jam

A loose term for a stoppage caused by an object, usually a round or cartridge case, becoming jammed or fouled inside a weapon.

Joule (J)

A unit of energy. (Newton metre).

K

KE (Kinetic energy)

The energy of a moving object. Mathematically this can be expressed as $\frac{1}{2} MV^2$.

L

Lands

The ridges of the rifling inside the bore.

Lead

The gap between the front of the chamber and the commencement of rifling. Another term for it is 'free flight'.

LMG (Light machine gun)

    An automatic weapon which usually has a bipod support for the barrel. It is readily portable by one man, is often magazine fed and is capable of providing a high volume of accurate fire out to 800 to 1000m.

LSW (Light Support Weapon)

    It has the same performance as the LMG except it may well have a less rigorous volume of fire requirement.

    M

m

    metre.

Machine pistol

    See SMC.

Mega. (M)

    One thousand times.

Milli. (m)

    One thousandth of.

MMG (Medium machine gun)

    An automatic weapon which is normally mounted on a tripod. It is less portable than a LMG, is belt fed and is capable of providing a high sustained volume of accurate fire out to 2000m or more.

Momentum

    A measure possessed by a moving object. Mathematically expressed as mass x velocity.

MPI (Mean point of impact)

    The centre of a group.

Muzzle energy

    The kinetic energy of a bullet at the muzzle.

    N

Newton. (N)

    A unit of force (kilogram metre per second^2).

    O

Obturation

    Prevention of leakage of high pressure gas.

Open bolt

> Weapons which fire from an open bolt have the working parts to the rear and no round in the chamber when the trigger is pressed.

P

Primary extraction

> The initial rearward movement of the empty case.

Primer

> An easily initiated explosive. Normally found at the base of a cartridge. When initiated electrically or by percussion it sets off the main propellant charge.

R

Radial

> Along the radius.

Recoil

> The rearward impulse which occurs when a weapon is fired.

Repetition

> The firing of single shots each requiring a separate trigger operation.

Rifle

> A long barrelled small arm capable of firing single shots very accurately out to 600-1000m. The name is derived from the rifling in the bore.

Rifling

> The spiral grooves in the bore. They are used to spin, and so stabilise, bullets.

S

Sear

> A device often found in trigger mechanisms. It engages in the bent in the breech block or bolt and prevents the latter from moving forward.

Self loading

> The weapon is automatically reloaded on firing without the firer having to take any action.

Semi automatic

> A weapon which is automatically reloaded but can only fire single shots.

SF

Sustained fire.

SLR

Self loading rifle

Small arm

A weapon which is generally manportable and fires a flat trajectory projectile. Normally small arms are considered to have calibres less than 12.7 mm. The distinction between a small arm and a cannon is blurred.

SMC (Sub machine carbine)

An alternative name for an SMG mostly used in the USA and Europe and synonymous with a machine pistol.

SMG (Sub machine gun)

A short barrelled automatic magazine fed weapon which is used for close quarter battle. It has poor accuracy at all but very short range but is light, simple to operate and cheap to manufacture.

T

Trajectory

The flight path of a projectile between the muzzle and the target.

Trunnion pull

Impulse on the mountings of a weapon.

Twist of rifling

The length required for one complete turn of a spiral groove in the bore.

V

VMMG

Vehicle mounted machine gun.

W

Watt

A unit of power. (Joules/sec).

Wear

Mechanical abbrasion which removes layers of metal.

Y

Yaw

The angle between the nose of the projectile and the line of its trajectory.

Z

Zeroing

The adjustment of the sights of a weapon so as to move the MPI of a group to the correct position in relation to the point of aim.

Bibliography

"Handbook of Infantry Weapons", RMCS, Shrivenham, 1971.

Lawton B, "Heat Transfer in Gun Barrels", RMCS, Shrivenham, 1977.

"Text Book of Small Arms", HMSO, London, 1929.

"Text Book of Small Arms", (provisional draft), 1953.

Index